W9-BZS-020

The Civilized Engineer

Also by Samuel C. Florman

Engineering and the Liberal Arts: A Technologist's Guide to History, Literature, Philosophy, Art, and Music (1968)

The Existential Pleasures of Engineering (1976)

Blaming Technology: The Irrational Search for Scapegoats (1981)

The Civilized Engineer

Samuel C. Florman

St. Martin's Press New York

Chapter 1 is based upon an article that appeared in the Winter 1985–1986 issue of *The American Scholar.*

Portions of Chapters 2, 9, through 16, 20, and 21 are adapted (with substantial additions and modifications) from articles that appeared in *Technology Review.*

Portions of Chapters 6, 7, and 8 appeared in IEEE *Spectrum, Business and Professional Ethics Journal,* and *Civil Engineering.*

Chapter 5 is based upon the Ralph Coats Roe Lecture delivered to the American Society of Mechanical Engineers in 1982 and subsequently published in *Mechanical Engineering.*

A portion of Chapter 17 appeared in *Science 86.*

Chapter 23 appeared in *Harper's.*

Design by Paolo Pepe

Library of Congress Cataloging-in-Publication Data

Florman, Samuel C.
The civilized engineer.

1. Engineering. 2. Technology. I. Title.
TA145.F55 1987 620 86-26183
ISBN 0-312-00114-2

First Edition
10 9 8 7 6 5 4 3 2 1

For Peter Cooper friends

Contents

Introduction

We live in the age of high tech. The signs are everywhere, from Hollywood to Wall Street, from Washington D.C. to fabled Silicon Valley. Newsstands display stacks of science and computer magazines. Windows of electronics stores are piled high with new products. Foreign policy experts attempt to come to terms with "Star Wars." Engineering stands at center stage.

Employment of engineers has been increasing at a pace nearly double that of other professions and three times as fast as that of the overall work force. More than 100,000 new students are crowding into American engineering schools each year, double the number of a decade ago. One of every five male college freshmen says he would like to be an engineer.[1] (The figure is much lower for women, but has risen dramatically in a generation.)

An important feature of this animated scene is the computer, trigger of the vaunted information explosion, herald of the era of communications and robotics. But the computer is only part of the story. The recent engineering renaissance stems in no small measure from a series of unpleasant shocks with which American society has been confronted: a war lost in Vietnam, a brief but worrisome energy crisis, and the growth of economic competition—fierce, bone-chilling competition—from the Orient and Western Europe. These unsettling developments, added to the long-standing pressures of cold war and nuclear stalemate, evoked first a sense of malaise and then a fierce determination. Americans have decided that they *will not* be buried under the sands of time. If, as events seem to indicate, our future depends upon technological prowess, then technologically adept we will be. Thus engineering becomes literally the key to survival. As a young man says in a TV commercial for the United States Army: "Technology is taking over the world. You keep up with it or you're going to be left behind. I don't intend to get left behind." This pledge becomes a statement of national purpose and also serves as a guide for many earnest young people.

The central role of technology in society has been an ongoing theme in American life, expressed for most of our history in an optimistic, often naïve, faith in "progress." But after World War II the proliferation of technical marvels took place against a counterpoint of doubt and disappointment. The 1950s brought with them anxiety about nuclear weapons and, after Sputnik, concern that the nation was losing its preeminence in science and engineering. The 1960s were marked by youth rebellions against, among other things, the "materialism" of American culture. With the 1970s came the environmental crisis and, for many engineers, a crisis of confidence.

Yet here we are in the latter part of the 1980s and technology is "in" again. Protests and concerns have not disappeared, but the public's attitude toward engineering is palpably upbeat. Even the most dreadful technological disasters—the escape of lethal gas in Bhopal, the loss of the shuttle *Challenger,* the reactor explosion at Chernobyl—are greeted with sorrow and expressions of resolve rather than with the virulent hostility that was typical of only a few years ago. People direct their ire at the failure of particular individuals and groups—usually politicians, administrators, or corporate executives—rather than at the concept of technical progress, which according to polls, they heartily endorse.[2] Proponents of conflicting political ideologies increasingly agree that hopes for peace, justice, and a more noble society are tied to the prospects of technological advance.

For those of us who are members of the engineering profession, the improvement in our circumstances is pleasant, to be sure, but somewhat disorienting. "Zoom!" as humorist Russell Baker has put it. "You have to be fast nowadays. Things go by at such a rate! . . . New sensations . . . new revolutionary ideas. They come by at a thousand miles a minute. Zoom!"

As engineers, we have been trained to cope with the unexpected, and if the unexpected is a turn for the better, then we certainly ought to be able to take it in stride. Yet life is subtle and capricious, and in some ways better can be worse, or in any event, not as much better as first appearances indicate. The improved status of engineering is clearly related to the conservative mood that has taken hold in our society, a mood generally identified with the advent of the Reagan Administration in early 1981. This is both good and bad. Since the new conservatism is *realistic* it is dear to the heart of every engineer. Once and for all, sensible people have agreed that there is no free lunch; there are only difficult choices, options, and trade-offs. On the other hand, a preoccupation with realism and tough-mindedness can have adverse consequences. What, for example, is happening to the vision and idealism that once were central to the engineering enterprise?

It is good to know that knee-jerk opposition to worthy technological projects has abated. But it cannot be counted a plus when newly assertive self-styled "pragmatists" belittle the concerns of environmentalists and propose to dismantle regulatory agencies. I take little satisfaction in knowing that engineering enrollment has doubled when I learn that young people are entering the profession mainly because it promises steady employment. It is depressing to learn that college students in general are far less concerned than they used to be with helping others or developing a meaningful philosophy of life.[3] Survive, prevail, and, if possible, get rich—that seems to be the motto for the moment. A melancholy paradox: in its moment of ascendance, engineering is faced with the trivialization of its purpose and the debasement of its practice.

I sometimes fancy that I am a Roman engineer traveling to the East with conquering legions. We bring with us skill and organization. We create roads and aqueducts, marbled halls and tiled baths. We improve "living conditions." Yet in this fantasy I see strewn all about us the ruins of an earlier Hellenic civilization, traces of an art and architecture whose grace contrasts with our avowedly utilitarian works. What can we learn, I wonder—we who prize efficiency—from a culture that prized truth and beauty? Then, waking from my reverie, I ponder the grim fact that Greece, for all its art and philosophy, and Rome, for all its wealth and technology, both in the end toppled and fell. Perhaps a culture that weds competence to grace, and wisdom to know-how, would persevere and flourish where others have failed. Such a culture would have at its core a cadre of civilized engineers.

As engineering becomes increasingly central to the shaping of society, it is ever more important that engineers become introspective. Rather than merely revel in our technical successes, we should intensify our efforts to explore, define, and improve the philosophical foundations of our profession.

Where does the profession come from? What are its origins and its history? What are its underlying views, traditions, and purposes? How does engineering relate to the greater community? And how, ideally, ought engineers to be trained and motivated? These are some of the questions I would like to consider.

If this sounds an overly ambitious note for a slim book of essays about engineering, let us at least try to enrich the passing moments in speculative consideration of our technological society. Since I first started writing for publication more than twenty-five years ago, I have met many concerned people—in correspondence as well as in person—who believe that this is both an important and agreeable thing to do.

The Civilized Engineer

1

Concrete and Kafka: A Personal Overture

"I became an engineer."

Thus begins John Hersey's novel, *A Single Pebble,* in which the protagonist travels to pre-revolutionary China seeking a site for a dam along the Yangtze River. As he encounters a civilization little changed since the Middle Ages, the young man finds his faith in technology giving way to awe and self-doubt.

I, too, became an engineer and have spent a number of years thinking about, as well as practicing, this much misunderstood profession, albeit in less dramatic settings than the chasms of the Yangtze, and with less discouraging conclusions than Mr. Hersey's.

How does one decide to become an engineer? I made the decision in 1942 during my senior year at the Fieldston School, a sylvan campus in the Riverdale section of the Bronx, forty-five minutes by subway and footpath from where I lived in Manhattan. The idea had occurred to me earlier—especially during several visits to the 1939 World's Fair—but I was far from being the stereotypical engineer. I did not, for example, build radios, assemble models, or fiddle with car engines. Living in the city, I had no access to cars, and when some mechanical device failed in our apartment, my parents called on the building superintendent. Like my fellow students at Fieldston, I read a lot of books and wrote a lot of papers. My favorite subject by far was English, particularly a senior seminar in which we reviewed great Western literature from Aeschylus to James Joyce. Nevertheless, I did my best work in mathematics, and was gently urged by several of my teachers to consider a career in science.

There were no "two cultures" in those days and I can recall no division between students of different sorts of talents, rather mutual respect and a shared appreciation of achievement. If this sounds idyllic, well it was. Not that we lived in a state of constant elation—we were teenagers, after all—but academically the place was heaven. We knew we would follow many different career paths: the world seemed incredibly open and full of possibilities—in the arts, the sciences, and the professions. Business, however, we regarded with a scorn compounded of intellectual elitism and post-1930s radicalism. Ironically, the fathers who paid our not inconsiderable tuition were mostly hard-working small businessmen.

Of the acceptable career alternatives, science, medicine, and engineering were considered more or less on a par with law, journalism, and the arts. Excellence is what counted; our class had an abundance of it and our expectations were high. We were not surprised in later years when the most accomplished student in the class studied physics at Harvard, got his doctorate there, and ended up at Los Alamos, any more than we were when the president of the student council became a nationally syndicated newspaper columnist, ofttimes called a pundit. We were a class full of potential pundits.

Although I wanted my life's work to be creative and stimulating, I was not totally oblivious to money. A part of my Depression-bred consciousness was concerned about some day being able to support a family. For all the appeal of mathematics and physics, it wasn't clear to me how one made a living in those fields. This was even more true of writing and the arts. Business, as I have said, was out of the question, and as for medicine, needles made me queasy. So I chose engineering. Engineers, from the little I knew, studied science and used their brains. They also got jobs and earned salaries. And, after a fashion, they were cultural heroes. The newsreels that I saw every weekend between two movies at Loews 83rd Street often featured the dedication of a new TVA dam or some other impressive public work. There was much cutting of ribbons and drinking of toasts, each event celebrating a counterattack against rural dust bowls or urban slums. And when the movies themselves depicted engineers—usually in the B film, to be sure—they were stalwart men in high-laced boots engaged in heroic endeavors such as building railroads or prospecting for oil. Intellectually challenging, financially sensible, and withal a touch of romance and adventure—engineering seemed like an ideal calling.

I had never heard it suggested that engineers were lower-middle-class, eccentric, or uncultivated—today I believe the epithet is "nerd"—nor did it occur to me that anybody held such opinions. The only sour note was sounded by an uncle who observed that instead of wanting to *be* an engineer I should aim to be someone who *hires* engineers, thus implying that I was about to join an exploitable sub-class. The remark enraged my father, who had no clear idea of what engineers did but was proud to have a son who was going to enter a profession.

When it came time to select a college, I naturally thought about M.I.T. Two of my engineering-bound classmates went to that august institution and never regretted their choice. But there was something about the huge labyrinth of laboratories that made my spirit sink, and still does in spite of all the good things I know about the place. Instead, I chose Dartmouth College, whose beautiful New Hampshire campus captured my heart.

I had only the vaguest idea of how one went about getting an engineering education. According to the Dartmouth catalog it seemed that I would "go to college," earn a Bachelor of Arts degree while majoring in the sciences, and then pursue an engineering degree in graduate school. This is how General Sylvanus Thayer thought it ought to be when, in 1867, he gave Dartmouth $40,000 for the purpose of establishing the Thayer School of Engineering. As Superintendent of the U. S. Military

Academy from 1817 to 1833, the general had overseen the development of that institution into a distinguished school of applied science, and in his later years he decided to endow a graduate school of engineering at a liberal arts college. He believed that before embarking upon professional training one ought to become "a gentleman." The Thayer School's two-year program originally was designed to follow a full four-year undergraduate education, but in 1893 a five-year program was devised combining the senior year of college with the first year of engineering school. That program endures to this day.

In most of the nation, however, engineering education evolved along different lines. The technical institutes, and later the land grant colleges, developed four-year programs that carried students directly from high school into engineering studies. This effectively did away with the concept of a liberally educated engineer, although the accrediting arm of the profession eventually required that an engineering curriculum have a minimum 12½ percent liberal arts component.

Of all this I was blissfully unaware as I arrived in Hanover, New Hampshire, in July of 1942. (A year-round program had been instituted because of the war.) I embarked on a typical course of study: English literature and French, sociology and economics, psychology and political science. As a pre-engineering student I also took mathematics, physics, and chemistry, and two other subjects—then required but long since discarded—graphics and surveying. I will not argue that these courses deserved to maintain their place in the curricula of higher education, but I recall vividly the delights of T-square, triangle, and india ink, and the thrill of carrying a transit through the autumn woods. These sorties into the tangible world, combined with the abstract fancies of mathematics and the sciences, reinforced my conviction that I was headed toward the best of all possible careers.

I was barely into my sophomore year when, almost imperceptibly, I began to undergo a metamorphosis. As if under a spell, I became increasingly absorbed in my technical, pre-professional studies. Looking back, I find it difficult to explain what happened, although since many of the hundred thousand-plus Americans who enter engineering each year go through the same experience, how extraordinary can it be? All I know is that the liberal arts began to pale and seem trivial, even annoying. Mathematical formulas took on the quality of fun-filled games, and the physical world became an enchanted kingdom whose every secret seemed worth exploring. Also I began to think of courses in terms of how they would help me become a better engineer, more thoroughly grounded in the sciences, more perceptive and quick-witted, and—let us face facts—more desirable to some future employer. Despite the educational advantages I had enjoyed in high school, and notwithstanding the proclaimed policy of the liberal arts college I was attending, I came down with a bad case of vocationalism. I lost interest in becoming an educated person—the "gentleman" envisioned by Sylvanus Thayer. I wanted to become an engineer.

Could an inspiring humanities professor have prevented this from happening? I like to think so. Surely the situation was not helped by a freshman English course devoted mostly to the painstaking dissection of *Lord Jim,* nor by the introductory social science courses which were informative but deadly dull. More exciting teachers and better-planned classes might have made the difference, but it is common knowledge that when one is embarked on an affair of the heart, the most prudent counsel, even skillfully presented, often falls on deaf ears. And there can be no doubt but that my feeling about engineering was not altogether different from falling in love.

As it happened, my most exciting professors were mathematicians. I recall winning a prize in a mathematics competition—second prize to be exact—and being invited along with the other winners to dinner at the home of the department head. After dining, we sat in the living room sipping brandy and listening to recordings of Mozart sonatas. Although at the time my musical taste ran more to Glenn Miller and Artie Shaw, I found the experience extremely agreeable. I associated my euphoria with the delights of mathematics, not giving adequate credit, I now believe, to Mozart and the winning of prizes, to say nothing of brandy.

Along with my commitment to mathematics and science, I developed a taste for extracurricular activities that I can only characterize as anti-intellectual. I and my pre-engineering fellows spent our leisure hours attending movies and sporting events. Occasionally we hitchhiked to Smith College and looked for girls. When lectures, concerts, and plays were offered on campus, it seemed natural that they be attended by other students, those increasingly strange young men who had decided to major in history, philosophy, or literature. One of my dormitory mates was enrolled in a special course with Robert Frost, who was at the time poet-in-residence at the college. Several times this friend invited me to join him for an evening of readings and discussion with the noted poet, but I was always too busy writing up my laboratory experiments, or else committed to a party at some local tavern. Today I cannot believe—*simply cannot believe*—that I never even saw Robert Frost, much less spent an evening with him when I had the chance.

Shortly after entering college, I had enlisted in the Navy V-12 program on campus, and at the end of each term there was a period of uncertainty while we recruits waited to hear what the government had planned for us. After a year, we were called to active duty, but this merely meant putting on a uniform and learning how to march in formation. Those of us who were heading for engineering were encouraged to continue our studies. After twenty-four months of nonstop schooling, I accumulated three years worth of credits and was ready to enter the professional phase of my education.

I had by this time resolved to become a "civil" engineer. The term was coined in mid-eighteenth-century England by John

Smeaton, builder of the Eddystone lighthouse, who wanted to demonstrate that his work had no military implication (which is ironic in view of the fact that most military engineers subsequently have been trained in civil engineering). Civil engineers design and construct buildings, dams, and bridges; towers, docks and tunnels—structures of all sorts. Civil engineering also encompasses highways, railroads, and airports, along with water supply and sewage disposal. In short, civil engineering is basic and of the earth, historically—along with mining—the root of all engineering. In the eighteenth century the development of the steam engine led to a new specialty called mechanical engineering, and each major technological advance has brought with it a fresh division of the profession: electrical, chemical, aeronautical, petroleum, computer, and so forth. I make this digression into the self-evident only because so many otherwise well-informed people keep asking me what it is that engineers *do.* Every technological product has to be designed and its fabrication overseen, and that is what engineers do. They occupy the vast middle spectrum between theoretical scientists and subprofessional technicians.

Buildings are usually planned by architects, but engineers design the structural and mechanical components within them, and civil engineers often oversee the actual construction process. These overseers are sometimes called construction engineers, and this is what I have become—more a business manager, I suppose, than a creative spirit, more a master builder than a man of science, yet still a member of the engineering family.

I have long forgotten most of the theorems that I learned in engineering school, but I recall vividly the nature, the "feel" of that learning. Like all engineers, I took basic courses in electricity, fluid mechanics, metallurgy, and thermodynamics (the study of heat and energy, particularly the workings of internal combustion engines, air conditioners, and the like). As a civil engineer, I took a series of courses in "structures," learning how to design beams, walls, slabs, and trusses. Then there were the more specialized studies: highways, water supply, and sanitation. In all of this there was a good amount of "hands-on" work. We poured concrete, cured it, and tested it to failure, analyzed the behavior of water in pipes and over weirs, and experimented with a variety of motors and generators. Occasionally we ventured out into the field, visiting construction jobs and sewage treatment plants, or —a great favorite—measuring the flow of a river while perched above it in a tiny hand-operated cable car. The theoretical work was difficult—some of it exceedingly so—but the physical *doing* made it seem worthwhile.

Nowadays engineering education is much more "scientific" than it used to be. In addition to the subjects that were taught in the 1940s, a contemporary curriculum will include computing and information processing, probability and statistics, systems, optimization, and control theory, even system dynamics (policy design and analysis based on feedback principles and computer simulation). Much of the so-called "shop work" has fallen by the wayside, relegated to students who take two-year technician courses or four-year engineering technology programs. The change came about in the 1950s, particularly in the aftershock of Sputnik. Also, the growth of new disciplines has meant that there is simply more material to learn and so less time for knocking about in overalls or muddy boots. This has been inevitable, appropriate, and a darned shame.

In the spring of 1945, the Navy decided to call in my debt, so to speak, and I was ordered to officers training school. After a few weeks of shooting guns, large and small, and studying semaphor code and naval etiquette, I was commissioned an ensign in the Civil Engineer Corps and sent to serve with the Seabees. I arrived in the Philippines just as the war was ending and went with the 29th Construction Battalion to occupy Truk, an atoll in the Caroline Islands that had been bypassed during the westward counteroffensive in the Pacific.

It was a pleasant enough life for a young would-be engineer.

During the day we worked on construction projects, repaving the airstrip with fresh coral dredged from the sea, erecting Quonset huts, and building roads and a water supply system. In the evenings we drank beer, played cards, and talked—mostly about our work, baseball, and girls. Also—and I could not remember this happening since early childhood—I found myself with long periods of spare time. As the tropical sun sank into the sea behind implausible palm trees, it was impossible not to become introspective.

Among tales of self-discovery and inner change, life on a desert island has an honored place. It helps, of course, if the island has a supply of books. Our battalion had a library of sorts, stocked mostly with the chunky, squarish paperbacks that were printed especially for the armed forces. To help pass the time, I started to read again. At first it was purely recreational: the mysteries of Erle Stanley Gardner, the historical novels of Thomas Costain and Samuel Shellabarger, the best-sellers of Lloyd Douglas and Irving Stone, and the outlandish burlesques of Max Shulman and H. Allen Smith. I had forgotten how much fun books could be, even ordinary unprepossessing ones. In addition to contemporary reprints, there were in our island library a number of Modern Library classics, and it was to these that I turned next. One evening I started to read *Crime and Punishment.* It was as if I had stepped through a looking glass and found myself back in my high school English seminar. Here once again were supercharged words and ideas and people and passions and questions of justice and the meaning of life. It was like high school and yet different because, though I was only four years older, I felt forty years more mature. On a Pacific island, thousands of miles from what we call civilization, the decades condense and the urge for meaning wells up in the young wanderer. At least this is what happened to me as I read *Crime and Punishment,* then *Madame Bovary, The Scarlet Letter, Pride and Prejudice, Fathers and Sons, Tess of the D'Urbervilles,* and other great works I cannot now recall.

I did not discuss with my fellow officers the books that I read and the thoughts that these books engendered. During the day, as the bulldozers roared and dust clouds rose, I immersed myself in the work as if nothing had changed. And in the evenings I idly chatted and joked like one of the guys. But I began to wonder why it was that we engineers, as a group, seemed to live in a world so far removed, intellectually and emotionally, from the ferment I was rediscovering in literature.

The only officer in the battalion who was not an engineer was the chaplain. One night he joined several of us in a game of cards, and between hands the conversation droned on in the usual way, which, as I have said, meant anecdotes about our work, the current baseball season, and girls. The chaplain tried to interject some thoughts about the United Nations and prospects for international order. We did not respond. He then tried other topics: the morality of nuclear weapons, the ethical responsibility of war criminals, the role of religion in the post-war world, and so forth, but each time we returned to our trivia. Finally, the chaplain slammed his cards down on the table, looked upward, and said in a loud voice, "Dear Lord, I know that I am unworthy, I confess that I have sinned, but why did you have to abandon me on this island with nobody for company but these boring engineers?"

I cannot say that from that moment on my life was totally changed, but it was an epiphany of sorts, a moment that I have never forgotten. I do not, needless to say, advocate wars or even universal military service. But there is much to be said for a forced interruption in a professional career. Every young engineer (or doctor, or physicist, or lawyer) would benefit, I am convinced, from a year on a desert island. The island should be well stocked with books and ideally should include among its inhabitants at least one outspoken chaplain.

* * *

When I returned to New York in the summer of 1946 I intended to look for a job. I had, after all, received a degree from Dartmouth and was entitled to call myself an engineer. (It was not until a few years later that I took the examinations necessary to acquire a state professional engineer's license, but that document is not a prerequisite for practice except in special circumstances, and most American engineers never bother to get one.) Many of my friends, however, were returning to school to resume educations that had been interrupted by the war, and this tended to make me feel that I was still properly a student. Besides, I was only twenty-one years old, the GI Bill of Rights offered free tuition, and the hunger for books that had been aroused during my stay in the Pacific was far from satisfied. I thought things over for a few weeks, then one September morning took a bus uptown to Columbia University and impetuously signed up for a master's program in English literature. My parents reacted to this as if I were manifesting some form of battle fatigue—without having been in a battle—and they were only partly placated by my assurances that I still planned to make my way as an engineer. The people at Columbia were likewise bemused. With only two semesters of English to my credit—freshman English at that—I was surely one of the least qualified candidates ever to knock at their door. But my intentions appeared honorable, and the mood of the day favored giving veterans a break.

I enrolled in four courses: American Literature Since 1870, taught by Lionel Trilling; Modern Drama, taught by Joseph Wood Krutch; The History of the English Novel, taught by a professor whose name I do not recall; and The Romantic Movement, taught by a professor whose name I do recall but will not record since he read his lectures in a monotone and almost managed to make Wordsworth, Shelley, and Keats seem lifeless. Aside from this one bad choice—for which I compensated by monitoring an undergraduate course in the Age of Reason—I found the classes totally absorbing. I remember with delight not only the lectures and the books we read, but also the arguments and small storms that constantly erupted. When Trilling announced in class one day that Henry James was a far greater novelist than John Steinbeck, and that anybody who didn't think so ought to reconsider his plans to teach English for a livelihood (most of my classmates were budding academics), the booing was loud and raucous. On the other hand, when Trilling entered the classroom after having testified in court on behalf of Edmund Wilson's *Memoirs of Hecate County*—which, strange as it may seem today, was almost proscribed for being pornographic—he was greeted with a standing ovation. When Krutch argued on behalf of trusting one's own judgment instead of kowtowing to so-called literary authorities, he was challenged by a student who said individual taste was suspect since "even monkeys know what they like"; to which Krutch replied, after a moment's thought, "I guess I prefer an honest monkey to a dishonest student." On another occasion, Krutch pointed to a political demonstration that was taking place on campus within view of the classroom window. "Who here," he asked, "believes that there is a radical solution to the discontents of mankind?" There then ensued a political debate followed by a discussion of the tragic view of life. This was totally different from anything I had ever experienced in an engineering classroom.

The entire campus seemed to seethe with excitement and I hardly knew where to look first. When Jacques Barzun was scheduled to lecture to the freshman Contemporary Civilization course, I wangled my way into the hall. A friend of mine recommended the music appreciation classes of the noted composer Douglas Moore, so when I was in the mood—which was often—I dropped in. It was a small class where I couldn't pass unnoticed, but Mr. Moore was a gentle man who said I was welcome. To this day, I remember how he analyzed with us the final movement of Mozart's 41st Symphony, the four themes intertwining in heavenly combinations. Paul Henry Lang, a well-known musicologist and critic, also was on the faculty and for a while I added one of his classes to my itinerant schedule. One day he confessed that he had begun to write a book on the aesthetic theory of opera only to give it up in dismay. Opera is beautiful and soul-stirring, he said, but its formal rules are undefinable. It simply is what it is. This is the way I felt about my experience at Columbia. I was no closer to Truth than I had ever been, but the overall experience was as thrilling as a Verdi duet.

In the evenings, after I had finished the required reading in Dickens, Ibsen, and Faulkner, I turned hungrily to the classics of earlier times, Homer, Shakespeare, Dante, and then, for a nightcap, read *The Partisan Review,* where I expected to learn what the heavy thinkers of the day were up to. On social outings, I started frequenting Carnegie Hall instead of the hotel ballrooms and Fifty-second Street jazz joints that had previously been my favorites. About the only habit that was not affected by my galloping intellectualism was the way I approached *The New York Times* on Sunday: I still started with the sports section—as I do to this day.

In order to qualify for a master's degree, I was required to write a thesis, and here I wanted to pick a topic that was all-embracing, not some hidden facet of an obscure author, but something "hot" that would reveal to me the essence of the contemporary cultural scene. With some trepidation, but determined to get advice from the highest possible source, I took the problem to Lionel Trilling. He was not enchanted by my concept, saying that it lacked subtlety, but then, on second thought, the question seemed to amuse him. After a moment's reflection, he suggested Franz Kafka. Kafka was culturally hot, he said, and getting hotter. This idea presented two problems: first, I had never read Kafka and had scarcely heard of him; second, and more serious, my selected area of interest—to which my thesis had to relate—was American literature. Well, said Trilling, why not American criticism of Franz Kafka? Why not, indeed? It was an inspired choice, for not only did I immerse myself in the anguished fantasies of this quintessential twentieth-century writer, but in tracking down his reviewers I came face to face with Marxism, Freudianism, existentialism, and all the other *isms* of the day. This excursion into the world of literary criticism showed me that intellectuals can be as petty, inconsistent, and plain foolish as anybody else, a fact that the awe-struck student of literature—and that is what I was—tends to overlook.

I had anticipated that the thesis would be a chore, but it turned out to be hard work, which is different and better. I enjoyed the research, delighted in the reading, and much to my surprise, relished the writing. I don't suppose that anybody has read that opus, or ever will, other than the kindly assistant professor whose job it was to do so, but a more rewarding task I have never undertaken. There were times when I found myself beginning to envy the men and women whose life's work is reading, researching, writing, and teaching, who traffic in words and ideas instead of goods.

Yet when the academic year ended, I said farewell to the campus on Morningside Heights the way one leaves an enchanting foreign land to return home. It never occurred to me that I might stay, become a scholar, and change the direction of my life. For one thing, it was high time to start earning money. Also, I was committed to engineering, I was trained, and deep down I believed that I was better suited to building than to reading and writing. In fact, deep down I had concluded that it was the main business of humankind to build, to be technologically creative. Literature, I felt in my bourgeois heart of hearts, was commentary upon life, while engineering was the stuff of life itself. What was the meaning of all those great novels except to explore the ways in which people work, trade, farm, war, earn, spend, cooperate, and compete? (Also love and hate and create art, to be sure, but only within the framework of an economically viable society.) What was the bitter joke with which Kafka wrestled during his short, sad life except that the state of grace that ever eludes the artist is found in the daily round of ordinary affairs?

As much as I was enchanted with literature and the life of the mind, I had come to resent the condescending attitude that some intellectuals evinced toward "materialism" and the technological impulse. It was all very well for Plato to say that thinking is better than doing, but it seemed ironic when one considered that the glory of Athens rested upon the marvels of Greek technology. (My resentment of Plato and his intellectual descendants was counterbalanced by my affection for Homer, whose literary world was rich with the feel of metals, woods, and fabrics, and whose robust characters took delight in buildings and ships, in objects designed, manufactured, used, given, admired, and savored.)

Shortly after leaving Columbia, I got my first honest-to-goodness job. As aide and lackey to a housebuilder on Long Island, I could not consider the work professionally exalted. But when I started to measure and calculate, to set up my transit and wave directions to the backhoe operator, I knew I was in a place where I belonged. Eventually my career took me into an office, and today I visit construction sites less often than I would like. But when the cranes lift steel beams to dizzying heights, or when the wet concrete flows into huge wooden forms as the laborers yell and bang their shovels against the chutes, I say to myself, as did Robert Louis Stevenson when he came upon a scene of railroad construction on the western American plains, "If it be romance, if it be contrast, if it be heroism we require, what was Troy to this?"[1]

I have had no reason to regret the career choice I made, somewhat cavalierly, in high school. If life has not been one grand adventure interspersed with ribbon-cutting ceremonies like those shown on the screen at Loews 83rd Street, well, it is common knowledge that life isn't the movies. Given real choices in the real world, I think that engineering ranks high in providing psychic rewards. Although I have a special affection for civil engineering, I recognize the appeal of the other branches of the profession, and vicariously enjoy the work of my more "scientific" fellows, for example, those on the frontiers of electronics.

Yet for all the satisfactions inherent in my work, I thank the fates that I had my time with the humanities and that I do not, like so many engineers in mid-career, lament lost opportunities in academe. Mark Van Doren overstated the case, but not by much, when he wrote of the happiness that comes to the student of the liberal arts:

> That happiness consists in the possession of his own powers, and in the sense that he has done all he could to avoid the bewilderment of one who suspects he has missed the main thing. There is no happiness like this.[2]

No one person can be all things at once. It is unrealistic to expect engineers to be artists or poets or literary scholars, and vice versa. Civilization flourishes because specialists rely upon one another to perform particular tasks. By nature and inclination, engineers are different from poets and artists. Having said that, I still think it regrettable that so many engineers arrive at the gates of their profession without having experienced much of what the world has to offer.

Occasionally I try to sell my personal brand of salvation to young engineers, but I do not delude myself about the extent of my success. If I try to speak of my own experience, some smart whippersnapper will tell me what I know only too well, that it was something of a fluke. Not every engineer can go to a private high school and a liberal arts college, then spend a year on an island in the Pacific followed by another year studying the humanities at government expense.

Yet who can predict the future? When the present frenzy over computers and technical training has run its course, the inevitable counterreaction will set in. Engineers, being bright, curious, and pragmatic, may well conceive new patterns for their education and their careers. When older engineers get together they invariably agree that immediately after graduating from college they wished they had taken more technical courses. Ten years later, advancing along career paths, they wished they had learned more about business and economics. Ten years again, in their forties, thinking about the nature of leadership and musing about the meaning of life, they regretted not having studied literature, history, and philosophy. This pattern has become something of a cliché, confirmed by studies and polls.

Can it ever change? Of course. At least, "Hope springs eternal in the human breast." The line is by Alexander Pope, who, as I learned at Columbia in 1947, was an eighteenth-century English poet with a penchant for quotable couplets. The sentiment, however, is very much in tune with the engineering temperament.

2

The Existential Pleasures of Engineering—A Generation Later

One evening in the fall of 1968 I received a telephone call from a member of the New York Academy of Sciences who was trying to round up speakers for the Engineering Division's monthly meetings. Usually, these get-togethers focus on technical topics, but he thought I might provide some variety by considering engineering from a broad philosophical point of view. My credentials, such as they were, consisted of a number of speculative essays that had appeared in various professional journals and a recently published book entitled *Engineering and the Liberal Arts.* I accepted the invitation and then began to wonder what in the world I was going to say.

Nineteen sixty-eight was not a good year for discussing the philosophy of engineering. All over the nation young people were in rebellion. The so-called "flower children" were attempting to come to terms with their inner selves, seeking to find expression for an ineffable longing they said was being thwarted by a materialistic society. *The Greening of America* had not yet been written, but its message was being broadcast in wistful song lyrics and in decidedly unwistful, occasionally frightening, student demonstrations.

I asked myself what it was that these young people were seeking. Love and friendship? Yes, we all seek these first and count ourselves blessed when we find them. Peace and justice in the world? Of course. But as for those other amorphous yearnings of the sixties, they seemed to me pathetic and unfruitful. The hippies disavowed "progress" and advocated "dropping out." But would that not lead inevitably to a worsening of the world's problems? And for themselves as individuals, would not such a course lead to a dead end, to lethargy and terrible boredom? It occurred to me that these young rebels were blaming their psychic anguish on a mysterious demon called Technology when it was really the misuse of technology by imperfect humans that was involved. Is not constructive work a cure for such discontent rather than a cause? Does not the technologically creative person in his or her daily tasks find, almost incidentally, the very satisfaction that the hippie searches for in vain? With these thoughts in mind, I delivered a talk entitled "The Existential Pleasures of Engineering." When, some months later, it was printed in the *Transactions* of the Academy, I received letters from engineers all over the world. "Yes," they said in essence, "we, too, have strong, positive philosophical feelings about the work we do." Encouraged by this response, I eventually expanded that talk into a book, published in 1976. (Some nice things were said about that book but I cannot claim that everybody liked it. An English professor writing in *The Chronicle of Higher Education* said that it was good mostly for cheering up engineers after a hard day before the Environmental Protection Agency. The two cultures live on!)

Existential. Quite a word. A lot of people asked me why I couldn't have used a less pretentious-sounding adjective. In fact, I had tried, but I couldn't find a better way to say what I wanted to say.

Existential feelings are those mental and emotional experiences that arise out of the depths of our innermost being—our intuitions, our basic impulses, what we feel in our heart (which, as Pascal said, has reasons that reason cannot know), what we feel in our bones, or in our gut—what Orientals mean when they speak of thinking with the belly.

I think that engineering is what human beings, deep down, want to do. Not the only thing, but one of the most basic and satisfying things. Engineering is an activity that is fulfilling—*existentially.*

There are biological reasons for this, of course. Both genetically and culturally the engineering instinct has been nurtured within us. To be human is to be technological. When we are being technological we are being human—we are expressing the age-old desire of the tribe to survive and prosper.

Within our hearts, however, we sense that more than biology is involved. Engineering responds to impulses that go beyond mere survival: a craving for variety and new possibilities, a feeling for proportion—for beauty—that we share with the artist. As engineers, we feel an urge to challenge nature—fighting storms, floods, earthquakes, and other life-threatening forces—yet also to work in harmony with nature, seeking understanding of soils, metals, and other basic materials of the earth. We partake of the wonders of the natural sciences and enter the pristine realm of mathematics. Our work contributes to the well-being of our fellow humans. There are religious implications in technology—a little bit of cathedral in everything we build.

My proposition is that the nature of engineering has been misconceived. Analysis and practical creativity do not preclude emotional fulfillment; they are pathways to such fulfillment. They do not "reduce" experience, as is so often claimed; they expand it. At the heart of engineering lies existential joy.

Almost two decades have passed since I first gave voice to these sentiments before a friendly, if slightly bemused, group in the elegant town house headquarters of the New York Academy of Sciences. In some ways it seems like only yesterday. But in some ways that evening seems to have taken place in another age.

We now live in the no-nonsense 1980s and the flower children are gone. If the truth be told, I miss them—not for their music, drugs, and languid ways, but for the challenge they presented to the rest of us. The young rebels of the sixties were dissatisfied with what America was becoming and, paradoxically, their very dissatisfaction was in the best American tradition. Their longing for a better society was a manifestation of New World utopianism. Their search for inner fulfillment was a poignant example of "the pursuit of happiness." They floundered, and in the end their movement failed, but their divine discontent is the same force that animates all worthy human endeavor, not least engineering. Out of the discontent of the sixties there grew a more humane technology, with more concern for aesthetics, safety, and environmental preservation. And among engineers there developed a keener appreciation of the moral importance and creative richness—the existential pleasures—of their profession.

Nowadays, however, we hear little of utopian dreams, more of concern about survival. In 1968 we were still talking in terms of a Great Society and looking forward to the first manned landing on the moon. Today we are cutting back on social services, worrying about competition from the Japanese, and thinking of space largely in terms of military application. In 1968 a study showed that young people chose engineering as a profession mostly because it promised "interesting work";[1] today the choice is made increasingly because of "job opportunities." The authors of a report in *Engineering Education* conclude that engineering is "not usually chosen from any love for the subject itself."[2] The proportion of college freshmen who say that "being very well off financially" is a very important goal has grown from 49 percent in 1969 to 65 percent in the 1980s. At the same time, those who value the goal of "developing a meaningful philosophy of life" has declined from 83 to 49 percent. Goals such as "helping others in difficulty" have been declining in importance, as well.[3] Half of all college students now major in career-oriented fields such as business, engineering, computer science, education, and nursing. While vocational enrollment increased, the humanities declined from 18 percent of total majors in 1969 to 5 percent in 1984. According to specialists who recruit technical professionals for employment, "The lucrative stock options and extravagant Christmas parties of high-technology companies now dominate the fantasies of student engineers."[4]

A survey of electrical engineers—the largest field of specialization within the profession—finds that the youngest EEs "have several attributes in common with what the media have called the 'me generation.' " It is perhaps unreasonable to expect that young engineers will be different from young people in other walks of life. Yet, as the poll-takers note, "it is nonetheless jarring to see signs that the new generation of electrical engineers—who hold the future of the profession in their hands—show less enthusiasm for the future of the profession than for satisfying their own immediate needs."[5]

I do not mean to suggest that everything has changed for the worse since 1968—I do not believe this at all. But I am concerned that the quality of the engineering experience may be corroded by the social changes that have occurred.

The rise of the computer has added to the problem. The phrase "hands on," which has an honored place in engineering tradition and used to refer to gasoline engines, electric motors, and concrete testing machines, now means running one's fingers over a keyboard and looking at numbers on a glowing screen. Many people attest to deep-felt, almost mystical, satisfactions in their work with computers. The computer engineers in Tracy Kidder's *The Soul of a New Machine* speak of "the golden moment," of "almost a chemical change" inside themselves when understanding dawns or the work goes well. "When it worked I'd get a little high." "It was the most incredible, soaring experience of my life." These statements reflect experiences comparable to the one that Rudolf Diesel tried to describe shortly before his death in 1913. He said that the conceptual breakthrough that led to the invention of his engine had filled him "with unutterable joy."[6] It is difficult to compare and evaluate such assertions, and I suppose that one ought not to try. "In the thick of active life," as Santayana said, "there is more need to stimulate fancy than to control it."[7] But whatever the benefits and satisfactions inherent in the so-called electronic era, I believe that the engineer's physical closeness to the natural world has diminished somewhat, and regrettably so. Of course, this may just be the civil engineer in me speaking. Happily, the loss of "earthiness" is somewhat offset by the enthusiasm of many young computer experts. What I really deplore is the waning of enthusiasm among would-be engineers, and the growth of cool, calculating self-interest.

But, the question might be posed, if the engineering experience is as great as you say it is, won't it inevitably bring out the best in engineers no matter what their reasons for entering the profession? Can't engineering be appreciated without regard to the public mood of the moment and without regard to the philosophical outlook of the individual? The response to this is yes, but only to a degree. The quality of an experience depends partly upon what the individual brings to it. I suppose that beavers "enjoy" the work they do, yet most of us aspire to more than animal-like satisfactions. Many of the experiences we relish in apparent spontaneity actually depend upon our pre-existing beliefs and intellectual awareness. An insensitive, undereducated, self-absorbed engineer will not derive as much from his work—nor give as much to his work—as a large-spirited, broadly cultured colleague.

Anne Morrow Lindbergh's short book, *Gift from the Sea,* was a tremendous success when it was published in 1955, mainly because it was a lovely, lucid work that came "from the heart." The author, after spending a few weeks on an island, mused philosophically about life—about simplicity, solitude, romantic love, middle age, the balance of physical, intellectual, and spiritual experience—her thoughts prompted by the discovery of various shells on the beach. Nothing, it seems, could be more elemental and less dependent upon intellect and culture. Yet in her meandering, personal speculations, Mrs. Lindbergh referred repeatedly, almost incidentally, to the great literary works of

Western civilization. She quoted Socrates: "May the outward and inward man be at one"; and Donne: "No man is an island"; and the Psalmist: "My cup runneth over"; as well as Rilke, Blake, Auden, Plotinus, St. Catherine, Virginia Woolf, William James, and others. She alluded to the story of Jason and the Golden Fleece, to Greek architecture, and to Mozart. The many thousands of readers who praised the book for its simplicity and artlessness were in fact responding to the sensibility of a highly educated and extremely cultured woman.

One need not be a classical scholar to find richness of experience in engineering. Technical creativity has its own inherent splendors, and they cannot help but appeal to even the most callous practitioner. The same recruiting experts who note the appeal of stock options and extravagant Christmas parties conclude that ultimately what most satisfies engineers is "complex, technical problem-solving and the opportunity to work with stimulating colleagues and to make a meaningful contribution to society."[8]

Nevertheless, without imagination, heightened awareness, moral sense, and some reference to the general culture, the engineering experience becomes less meaningful, less fulfilling than it should be. When getting a job is more discussed than the quality of life, I fear that the existential pleasures of engineering may be adversely affected. It is the obligation and privilege of each engineer to do everything possible to prevent this from happening.

3

Heritage: (1) From the Beginning

"Where do I come from?"

Every child asks the question, and most of us continue to ask it in various ways throughout our lives. It is a good and useful question. The more we study our origins, the more we are likely to learn about who and what we are and where we may be heading.

Of course, the living present has its prior claims. The engineering we do today needs no historical justification. But commitment to the present need not imply disregard for what has gone before. There is a natural rhythm in life that allows for the interplay of opposites. The hearty spirit welcomes reflection on the past as an interlude between concentrations on the moment —and as a source for dreams about the future.

As members of a pioneer society, Americans have generally been forward-looking. The excitement of each day has been so intense, and the promise of tomorrow so bright, that history has often seemed dim and irrelevant. Engineers in particular have been preoccupied with the moment and entranced with the future. But in recent years America's frontier enthusiasm has become tinged with anxiety; our society's prospects appear clouded. We hesitate. We stop to catch our breath and to reevaluate our course.

It has become fashionable to say that one should study the past in order to avoid repeating its mistakes. But this is a limited and excessively sober view. History also reveals to us the aspirations and accomplishments of other civilizations. As noted by the British historian, G. M. Trevelyan, "The ideals of no one epoch can in themselves be sufficient as an interpretation of life." Thus, beyond its instructive and cautionary function, history can—again in Trevelyan's words—"breed enthusiasm."[1] In these times of uncertainty what could be more welcome? For those of us who are engineers the moment seems right for a retrospective look at our profession. Where do we come from?

Strictly speaking, the engineering profession came into being in the eighteenth century when science was first applied to the solution of technical problems. But by adhering to such a standard—by insisting that without science there is no engineering—we would exclude from our family tree many geniuses, from the builders of the pyramids to the inventors of the steam engine, and this we refuse to do. For present purposes, let us set aside definitions and broaden our conception of engineering to include all technological creativity.

The engineering impulse developed in our anthropoid progenitors as they roamed across the plains of Africa and Asia several hundred thousand years ago. These creatures of nimble hands and large brains were not skilled builders like birds, beavers, and spiders. They had not followed an evolutionary road toward preprogrammed behavior. They were instinctive technologists, but of a very different sort. Their distinctive traits were intelligence, curiosity, and adaptability.

Hunting wild game for food, they undoubtedly wielded sticks and animal bones as clubs and used rocks as missiles. In order to remove skin and fur from the animals they killed, they seized upon stones with sharp edges and used them for cutting. In other words, they used found objects as tools. This required perceptiveness and understanding, but not quite the creativity that we identify with being human. Apes will pick up objects at hand to accomplish simple tasks. For example, baboons will roll rocks down a hill behind them when fleeing uphill from enemies, or use them as tools to crush scorpions, a favorite food. A chimpanzee in captivity is known to have fit two bamboo tubes together in order to reach a bunch of bananas outside his cage, and even to have sharpened a piece of wood with his teeth in order to fit it into a bamboo shoot for the same purpose. But while apes may perceive the solution to a visible problem, or even improvise a tool to meet a given situation, they appear incapable of shaping an implement for use in an imagined situation. Conceptual thought—the formation of abstract ideas—distinguishes human beings from even the cleverest ape.

Our ancestors learned to save those naturally formed implements that best suited their purposes, and eventually realized that they could duplicate the processes by which objects were turned into utensils. By purposefully breaking a stick or a bone, or—momentous act in the history of humankind—by chipping a stone, they could make tools for use when needed. The first individuals who thus engaged in creative design were the first engineers. They were also, according to some students of evolution, the first human beings.

The creative process must always remain something of a mystery, and its origins in pre-history even more so. We know that engineering reasoning consists of induction (observing, and drawing conclusions from one's perceptions) and deduction (logical reasoning from one general principle to another). Add to this a touch of what one might call inspiration and the basic recipe is complete.

Clearly, evolutionary forces played a role. A better spear meant a better dinner. A surer way of starting a fire meant an improved chance of surviving a cold night. Natural selection favored the clever and the adept. But this is merely one aspect of a process of awesome complexity.

We speak of Stone Age culture because of the stone tools that archeologists have uncovered in great numbers. The terminology, however, is somewhat misleading. Stones are relatively unaffected by the passage of time, whereas implements made of other materials—and there must have been many of wood, bone, and shell—are likely to decompose. We can infer from the shape of certain stone tools that they were used to make wooden objects. This, plus a few fossilized wood remains, plus a bit of common sense, convinces us that from early Paleolithic times humans made tools, not only out of stone, but out of whatever materials were available.

Slowly—over thousands and thousands of years—tool-making techniques evolved. If we visit a museum and look intently at the tools of so-called Stone Age cultures, we cannot help being impressed by the many examples of technological genius. Long before the beginning of recorded history, human beings were using a variety of ingenious implements: axes, knives, and spears; needles, harpoons, and barbed fish hooks; traps and snares; bows and arrows and spear throwers. They scraped animal skins for use as clothing and probably as tents. They ground black oxide of manganese and red and yellow oxides of iron, mixed them with fatty substances, and made paints with which, more than 20,000 years ago, they drew stunningly graceful figures on the walls of caves.

Because life was limited to a bare subsistence level, dependent upon food gathering and hunting, family and tribal groups were necessarily small. It is therefore unlikely that any specific members of these groups were designated as technologists: everybody must have been trained in the basic skills that were held in common and transmitted from one generation to the next. In such a situation, individuality is not encouraged; one does what one's parents did. Nevertheless, the hand-making of tools is a uniquely individual experience. It evokes a deep-seated "feel" for materials and a "knowledge" that is distributed throughout the body—as much in the fingers and eyes as in the brain. Out of craftsmanship there arises, occasionally and mysteriously, the engineering impulse. An individual will perceive ways in which tools or techniques can be modified. This perception becomes joined with the urge to try something new, to experiment and to improve, and suddenly the worker becomes a creator.

No matter how closely modern engineering becomes identified with science, no matter how electronic or theoretical its practice, no matter how much a group enterprise it may be, the end product of which seems remote and abstract, it can never be —should never be—severed from its origins in craftsmanship.

Craft also gives rise to an artistic impulse, and our ancestors very early demonstrated a desire to make utensils beautiful as well as useful. There is ample evidence that the urge to fashion objects, and to lend grace to their shapes, is deeply lodged in the human psyche. Even as we master methods of mass production and seem about to cover the earth with polyester fabrics and plastic utensils, we are witnessing an extraordinary rebirth of the ancient crafts. People take up woodworking, weaving, quilting, and pottery as if to demonstrate the existence of some primordial force. And even in the most utilitarian machine-made product we are apt to find the grace and proportion that in modern parlance we call "design." In a world where technology is too often viewed as an enemy of aesthetics, we do well to recall the common roots of craft and art.

Science—as distinct from engineering—was not a purposeful activity of primitive peoples. But success in hunting depended upon accurate observation of the habits of game. Gatherers of roots and berries had to learn about plants experimentally. Man's dependence upon nature led to his study of the seasons and eventually to his examination of the heavens. Ancient rings of stones, aligned with the movement of sun, moon, and stars, are evidence of a quest for scientific understanding going back almost as far as the beginnings of technology.

Among the peoples of pre-history, specialization was an unaffordable luxury. Nevertheless, we can discern in their works the origins of the separate career roles that characterize modern societies. The earliest division of labor was probably between the sexes, and the traditional view has been that men became hunters, warriors, and technologists, while women became child-rearers and homemakers. But if we look far enough back into pre-history, the picture is not quite that tidy. Surely the technologies of fiber, cloth, basketry, and pottery are closely linked with females. And if early man did indeed do most of the hunting, it is likely that early woman did most of the firetending, and perhaps much of the toolmaking as well. In the greatest technological revolutions of human history—the domestication of animals and the cultivation of plants—females played an important, some say decisive, role. And perhaps it was a woman, at the hearth, about 6,000 years ago, who noticed the accidental smelting of the common green ore, malachite, picked up a few bright globules of copper, and led humanity out of the Stone Age.

It is also interesting to note that most early technological progress was made in Africa and Asia. In Northern Europe, barbarians were still using stone axes 2,000 years after the manufacture of copper tools in Egypt. I say this is interesting because, during the Industrial Revolution, engineering came to be regarded as the special province of Northern European males, and this traditional stereotype was transplanted to the United States. If, however, we look to the deepest roots of engineering, the contrast is striking.

It is too bad in a way that humans could not have remained generalists, everyone a hunter and cook, a plier of craft and teacher of children, trader, artist, priest, engineer, and scientist. But technological progress depends upon a variety of skills and knowledge that are far beyond the capacity of any one individual. High civilization requires a high degree of specialization, and it was toward high civilization that the human journey appears always to have been directed. The curiosity and creativity that evolved and grew through the long years of human prehistory were too intense to be satisfied by the life that we now—sometimes wistfully and apologetically—call primitive. However, as long as there was destined to be a division of labor, engineers can be thankful that their share of the world's work is so interesting—so directly descended from the toolmaking that figured at the heart of the human adventure.

Prehistoric peoples progressed from subsistence hunting to the domestication of animals, and then to the cultivation of crops. Agriculture depended upon and generated new tools, such as the plough and the scythe, but it also brought about technology of a totally new dimension: vast construction works requiring the cooperative effort of many communities.

Around 4,000 B.C. the people who lived along the Nile started building irrigation canals to control the annual flooding of the river. These works grew and meshed until they stretched for hundreds of miles. Such enterprises required organization on a scale previously unknown to human society; they evoked a new institution, the national government. Government, of course, required and sustained scribes, priests, artists, warriors, and all the other specialists of civilized society.

Engineers of today can look back with pride and see how their predecessors made possible the breakthrough from primitive tribal life to the rich civilization of empires. The profession's association with public works, with communal projects of vast scope, is inscribed in the pages of history, and the title of engineer/bureaucrat can be worn as a badge of honor.

The hydraulic engineers in ancient Egypt—and in the other great river-valley civilizations—must have worked closely with the administrators and rulers of the time. To a certain extent they had to become administrators themselves, supplementing their technical know-how with managerial skills, taking on a very different role from that of independent artisan. It was not enough to say, let us run a ditch from here to there and put in a sluice gate of such and such size. The expert also had to consider how many workers were required and what tools and construction techniques should be used.

Yet the technical experts did not become the governors or the Pharaohs. To become a ruler in ancient times one had to be a warrior, or a leader of warriors. Or, like today, one had to be a politician—a promiser, threatener, and cajoler. Even then, charisma must have counted. The patient recorder of dates and flood levels, the clever calculator of volumes and pressures, the intuitive designer of weirs and dams, was not likely to be such a person. This is not to say that people trained as engineers today cannot enter politics. In fact, I will argue that the politics of today cries out for the participation of engineers. But it is clear that from the beginning of recorded history, technologists as a group were destined not to be rulers (and vice versa), though the association between the two castes was fated to be close and complex. They needed each other. Unfortunately for the technologist, while his contributions were important, often crucial, his role was inevitably subservient. If he was lucky, he was given a modicum of recognition, perhaps even respect, and a livelihood. The names of the great engineers of antiquity are mostly unknown. When a name *is* associated with a great work it usually belongs to a political bigwig. For the designers and the builders, the greatest reward must have been the esteem of colleagues, the knowledge that their contributions were important, and, of course, the satisfactions inherent in the creative process. Today the situation is not very much changed.

In their alliance with rulers, engineers have inevitably become involved in some unsavory enterprises. When the most skilled Egyptian builders were ordered to embark upon the construction of the pyramids, did they wince at the thought of devoting so much human effort to erecting the tombs of megalomaniacs? Did they think, why not instead build more ports, roads, schools, temples, and hospitals? Perhaps, but probably not. If pyramids were the going thing, then pyramids it would be. Engineers are often criticized for selling out to the ruling class. But it is only fair to assume that engineers saw the world around them pretty much as their fellow citizens did. They did not "sell out" or "go along"—they merely participated in the culture of their time.

* * *

By participating in the culture of their times, engineers inevitably became involved in the waging of war. If the earliest tools were related to hunting, it follows that weaponry is as old as engineering. Yet engineers did not make human beings warlike. People are not more aggressive with guns than they are with clubs. An argument can even be made, looking at the world since 1945, that the more destructive the weapons, the more cautious the citizenry. I will not attempt to make that argument here. I only wish to stress that in considering their own roots, engineers need not accept a disproportionate amount of the blame. On the other hand, when seeking glory for the great achievements of empire, engineers must humbly recognize the many dark pages of history for which they—along with their fellow citizens—share responsibility.

Perhaps the darkest pages of all are those devoted to slavery. The great structure of the ancient world could not have been built without the energy provided by uncounted thousands of human workers. Local populations were pressed into service by harsh rulers, receiving their "pay" mostly in food and protection. When prisoners of war became available, it seemed reasonable that they be used for the most onerous tasks. This must have begun casually on a small scale, but the practice grew and became formalized. The mines of antiquity were manned by slaves; so were the galleys. Slaves were used for construction, farming, and eventually, by those who could afford them, for household chores. Clearly, engineers considered slave gangs to be an acceptable source of power, a "given," the way a steam shovel or an electric drill was a "given" for engineers of later times. I say this, again, neither to blame engineers nor to absolve them. The times were what they were, and perhaps the good things that we identify with civilization could not have evolved without the widespread acceptance of this awful practice.

In the long run, of course, the effect of engineering has been toward abolishing slavery and ending human drudgery of all sorts. A central theme of the engineering story has been the search for mechanical efficiency and for sources of natural power, both of which can ease the need for toil. Long ago, animals were harnessed and used to carry loads, pull ploughs, and power grinding mills. By Roman times the waterwheel was in use, and in the early Middle Ages the windmill, known to Persia as early as A.D. 600, appeared in Europe. The steam engine was developed in the eighteenth century, internal combustion and electricity in the nineteenth, nuclear and solar power in the twentieth. The implicit goal of this continuing search has been freedom and dignity. In this sense, energy is seen to be something of a holy grail and engineers to be the knights errant of the modern world.

Although engineers of antiquity often played the role of slave-masters, in Greece, at the height of the Classical Age, events took an ironic turn. Plato and his cronies started saying that technological endeavor was properly the work of slaves, and, indeed, when Athens was at its peak around 400 B.C. many of the leading engineers *were* actually slaves. This social metamorphosis appears to have occurred rather abruptly.

The prominence achieved by the Greek city-states had earlier been founded upon a flourishing technology. The Greeks made great advances in the working of bronze and iron. They mined silver, pioneered in building tunnels, and excelled in constructing harbors. They became wealthy by being technologically creative. Shortly after 600 B.C., the Athenians authorized Solon, the chief magistrate of Athens, to initiate economic and constitutional reforms, and Plutarch tells us that as part of his plan to design a stable and prosperous society Solon "invested the crafts with honor." "At that time," says Plutarch, "work was not a disgrace, nor did the possession of a trade imply social inferiority."[2] A century and a half later, technology was still treated with high regard, as Sophocles made clear in the famous chorus from *Antigone:*

> Wonders are many, and none is more wonderful than man; the power that crosses the white sea, driven by the stormy south-wind, making a path under surges that threaten to engulf him; . . . turning the soil with the offspring of horses, as the ploughs go to and fro from year to year. . . . And speech, and windswift thought, and all the moods that mould a state, hath he taught himself; and how to flee the arrows of the frost, when 'tis hard lodging under the clear sky, and the arrows of the rushing rain; yea, he hath resource for all . . .[3]

But by Plato's time, in the first half of the next century (Plato died around 347 B.C.), the situation had changed radically. High culture had come to Athens, and with a vengeance. The ideal of the new Athenian citizen was to care for his body in the gymnasium, reason his way to Truth in the academy, gossip in the agora, and debate in the senate. Technology was not deemed worthy of a free man's time. Plato stated this emphatically in *Laws, Gorgias,* and elsewhere. His contemporary, Xenophon, summed up the argument this way:

> What are called the mechanical arts carry a social stigma and are rightly dishonored in our cities. For these arts damage the bodies of those who work at them or who act as overseers, by compelling them to a sedentary life and to an indoor life, and, in some cases, to spend the whole day by the fire. This physical degeneration results also in deterioration of the soul. Furthermore, the workers at these trades simply have not got the time to perform the offices of friendship or citizenship. Consequently they are looked upon as bad friends and bad patriots, and in some cities, especially the warlike ones, it is not legal for a citizen to ply a mechanical trade.[4]

Mathematics and pure science, on the other hand, were admired because they were not sullied by physical labor or "need." Even Archimedes, who lived several generations later (he died in 212 B.C.), felt obliged to make the distinction and so disparage his engineering achievements. Plutarch observed that "although his inventions had won for him a name and fame for superhuman sagacity, he would not consent to leave behind him any treatise on this subject, but regarding the work of an engineer and every art that ministers to the needs of life as ignoble and vulgar, he devoted his earnest efforts only to those studies the subtlety and charm of which are not affected by the claims of necessity."[5]

Historians are not agreed on the precise degree to which, by Plato's time, industrial techniques had passed into the hands of slaves. But for our purposes the details are not important. What counts is that Platonic scorn for technology has been transmitted along with "classical" education and has plagued engineers in every age. Long an article of faith at Oxford and Cambridge Universities, this prejudice became endemic among the ruling classes of England, and was eventually exported to America.

Of course, the outlook of classical antiquity does not begin and end with the age of Plato. It includes Sophocles, whose hymn to technology we have already noted, as well as Homer and the writers of the Old Testament, in whose works the emotional, spiritual, and practical components of technology are enthusiastically celebrated. But people have a way of seeing what they want to see, and the snobbish side of Plato appealed to aristocratic landowners concerned with maintaining positions of privilege. Wherever the sons of the wealthy were discouraged from studying engineering, the field was taken over by the less polished members of society, and reality came to mirror prejudice. As one might expect, engineers then engaged in a form of reverse snobbism, labeling non-technical learning as effete and without substance.

The social status of the engineer, first debated in ancient Greece, has remained in flux ever since. Sometimes engineers appear to be the "elite" of the laboring class, sometimes aloof professionals, and sometimes the ally of rulers. Since the growth of organized labor, engineers have argued whether, under certain circumstances, they should join unions or whether such commitment is, by definition, "non-professional."

Whatever their place in the social hierarchy from time to time, engineers are perennially the agents of technological progress and, as such, at the cutting edge of history. Where power rests in an entrenched nobility, particularly landowners, technology represents a threat: an opportunity for the underprivileged. Where factories spring up near castles and manor houses, where railroads and airports encroach on plantations, where radio and television become accessible to the masses, the foundations of hereditary class privilege inevitably erode. This has happened too often for us to doubt that engineers abet revolution, whether they intend it or not. On the other hand, where industrial societies have been established, engineering is likely to be viewed as an instrument of the established order.

Whatever the perplexities and paradoxes of the past, the time seems ripe for engineers to shed the Platonic stigma. With the burgeoning of electronics, engineering is less a dirty hands and muddy boots occupation than it used to be. This is not necessarily good—in some respects it is regrettable—but it removes one barrier to social acceptability. At the same time, our concept of what constitutes "aristocratic" behavior is less fastidious than it used to be. When traditional workclothes are transmuted into designer bluejeans, it becomes obvious that social underpinnings are shifting.

A generation ago most engineering students were young men from blue-collar families. Today the overwhelming majority are the children of college graduates, many the children of professionals, and they are exhibiting greater communication and social skills than ever before.[6] Also—and here the social consequences may be far reaching indeed—for the first time in recorded history an appreciable number of them are women.

I do not suggest that engineers should cultivate the affectations of social climbers. The profession's origin in handicraft and labor, its traditional lack of pretension, should be, if anything, a source of pride. I do believe, however, that Platonic snobbery toward technologists should be discredited once and for all and seen for what it is—the vestige of a slave society.

It is not appropriate, however, to think of Plato mainly as a disparager of engineering. The great philosopher's call to reason and his reverence for mathematics were landmarks on the road to modern science. Aristotle, who was a youth when Plato died, also combined disdain for technology with invaluable contributions to science. His textbooks in logic and his classification of the various branches of knowledge were to continue to bear fruit hundreds of years after he lived. While the philosophers of ancient Greece sought to place science on a higher plane than engineering, and succeeded in isolating the two endeavors from each other, they did help create a science that would some day be wedded to craftsmanship to become the modern profession of engineering. Thus the profession's severest detractors are also among its most inspired forefathers.

When the Romans strode onto the stage of history, they were less concerned about social niceties than were the Athenians of Plato's time. To them technology was not just a grubby necessity that made it possible for intellectuals to live comfortably; it was the stuff of life. The Romans continued the use of slaves—indeed they raised it to new heights of efficiency—but they were also willing to put their soldiers to work when they weren't busy fighting. Roman philosophers, to be sure, persisted in belittling technical achievements. Thus Seneca, after speaking of cloth mills and foundries, brilliant glass windows and enormous halls of polished marble, concluded: "All these are the inventions of the meanest slaves. Philosophy sits more loftily enthroned: she doesn't train the hand, but is instructress of the spirit."[7] Yet the Romans did not pay much attention to their philosophers. They lived by deeds rather than words. They taught a lesson about the importance of "infrastructure" from which we might well profit today. Their feats of building were nothing less than prodigious: 186,000 miles of good roads within the empire; ten aqueducts providing Rome with 220 million gallons of water daily; the dome of the Pantheon spanning more than 142 feet.

Our pride in the engineering accomplishments of the Romans is tempered by the recognition that many of their works were overblown and many of their customs barbaric. Their triumphal arches and blood-tainted coliseums pose an urgent question: How can a civilization retain a commitment to virtue and a feeling for aesthetics while at the same time pursuing wealth and material growth?

Rome fell, not because its technology failed, but because it collapsed politically, and most of all because it was overrun by other peoples of less wealth and more energy—another lesson for our time. In the Dark Ages that followed, it is impossible to believe that technical ingenuity flagged. But without social order and financial investment, inventive genius is of little avail, so in those times of turmoil, engineering progress was limited.

While migrating tribes criss-crossed the face of Europe, developing a patchwork of farms and fortresses and gathering in towns destined to become great cities, the center of engineering achievement shifted to the East. The Eastern Roman Empire lasted in Byzantium for a thousand years after the fall of Rome itself, and its domed churches with their stunning mosaics are testimony to a technology that thrived. The Arabs swept across North Africa starting in the sixth century and moved into Spain, where they remained for a thousand years. They established centers of civilization from Baghdad to Cordova and made great advances in mathematics (using "arabic" numbers), astronomy, navigation, and geography. They developed sophisticated waterwheels and windmills. In China, technologists made significant advances, notably in the production of silk, porcelain, gunpowder, and paper.

We Westerners—heirs of the Industrial Revolution and founders of the Space Age, the Atomic Age, the Electronic Age, et al—conveniently ignore the accomplishments of cultures about which we know shamefully little. Historians of technology are gradually helping us to see the gaps in our knowledge. However, even as we acknowledge the engineering achievements of the non-Western world, we cannot help noting that the autocratic Eastern cultures eventually stagnated, while the West—moving toward a boisterous individualism—became technically ever more creative.

When we speak of engineering in the Middle Ages, the first thing that comes to mind is cathedrals. In the presence of these spectacular structures we are inspired, and in thinking of the master builders who created them we feel a glowing pride of kinship. But the Christian Church, from its earliest days, was ambivalent in its attitude toward technology. The Gospels warned against giving too much thought to worldly goods. Yet honest labor was praised, and the monasteries became great centers of technical skill and knowledge. Eventually, the church became an important patron of the handicraft arts. The courts of feudal lords also provided support for the crafts, and in the towns independent freemen plied their technical trades. By 1300 an anonymous genius had invented the mechanical clock, a momentous achievement in the history of technology.

A remarkable degree of specialization developed. By the mid-fourteenth century, for example, there were more than twelve hundred "masters" plying no fewer than fifty different trades in the town of Nuremberg. Tanners, coopers, carpenters, dyers, braziers, locksmiths, tinsmiths, ironers, glaziers, potters, rope-makers, nail-makers—the list goes on and on, reminding us of the mechanical, metallurgical, and chemical skills that were being honed.

During this time, craftsmen banded together in guilds, organizations through which they attempted to improve and control the quality of their work—as well as uphold their business interests. The guilds were the precursors of contemporary professional societies and, in an age when there were no regulatory agencies, they served to protect the public as well as their members. Like all monopolistic enterprises they proved to be a mixed blessing. In recent years, through antitrust and other legal action, the American citizenry has sought to give government an oversight role in controlling the professions. Just how much self-regulation professionals should assume—and be permitted—is a question still subject to intense debate.

If the Middle Ages were not as technically backward as generally assumed, it follows that the Renaissance was not as much a rebirth as advertised. Nevertheless, the rediscovery and study of the writings of antiquity, plus the wider literacy brought about by the invention of the printing press, had consequences for technology as well as for literature and the arts. Manuals of instruction began to appear combining rule-of-thumb facts with mathematics and theoretical concepts. By the mid-fifteenth century, classical scholars like Leone Battista Alberti were applying the theories of the ancients to architecture and construction, while Filippo Brunelleschi, a goldsmith and builder of fortresses, was busy applying Euclid's work in his studies of perspective.

Of all the multi-talented men of that period, the most notable was the great Leonardo da Vinci. His career as both artist and engineer stands as one of the towering achievements of all civilization. Historians are fond of quoting his letter seeking employment at the court of Lodovico Sforza in 1482, a document in which Leonardo boasts mainly of his talents as a military engineer. He can build portable bridges, he claims, dewater an enemy's moat or undermine his fortress, make guns and mortars, contrive superior catapults, and even provide machinery for battles at sea. Only as a brief afterthought does he note that in time of peace "I believe I can give perfect satisfaction in architecture, building, sculpture, and painting."[8]

As noted earlier, engineering has always been identified with warfare. During the Renaissance, however, the association became increasingly formalized. It was during this period that the word *engineer* first came into common use, and the record shows clearly that the word's roots are military. As early as A.D. 200 the Latin word *ingenium*—meaning a clever thought or invention—was applied to a battering ram. By medieval times the men who devised catapults and other siege machinery were called "ingeniators," and during the Renaissance the modern words came into use—*ingenieur* in French, *ingeniere* in Italian, and *ingeniero* in Spanish. The English word appeared in the fourteenth century as *engynour,* in the fifteenth as *yngynore,* in the sixteenth as *ingener, inginer* and *enginer,* and finally in the seventeenth as *engineer.* Some engineer-writers have sought to stress the root in *ingenuity* rather than the association with *engine,* which they think sounds less professional. But since both *engine* and *ingenuity* stem from the same word—*ingenium*—not much is gained by the distinction.

The ingenious person has always been treated by his fellow citizens with a mixture of admiration and distrust. Both *engine* and *ingenuity* in their early forms were used to describe cunning and craftiness as well as skill. "For tis the sport," said Hamlet, "to have the enginer/Hoist with his own petar." A petar(d) was an explosive device used in sieges to break through gates and barricades, and many would agree with Hamlet that the ultimate in poetic justice is to see a military engineer blown up by his own infernal device, tricked by his trickery. A trace of this antagonism is directed toward the engineer in every age.

* * *

Out of the intellectual ferment of the Renaissance—rapidly and inevitably—grew modern science. The effect of science on engineering was at first negligible. In retrospect, however, we can see that the scientific revolution was closely linked with a technological revolution. As science evolved, engineering imperceptibly began to change from a craft into a profession.

4

Heritage: (2) In the Age of Science

In the early sixteenth century the Polish astronomer Nicolaus Copernicus declared that the center of the universe was the sun, not the earth, and as the seventeenth century dawned, Galileo Galilei, looking through his telescope, found confirming evidence. To the builders, miners, and mechanics who carried on the business of engineering in Galileo's day these cosmological concepts were of little interest. However, Galileo did not spend all his time looking at the night skies. He also conducted a wide variety of experiments with pendulums, projectiles, and falling bodies. In this he foreshadowed the work of Isaac Newton, who was born in the year of Galileo's death, 1642. Newton discovered the law of universal gravitation, which, in addition to explaining the motions of heavenly bodies, is the basis of what we engineers call mechanics. Among other accomplishments, he discovered that white light is composed of all the colors of the spectrum and—at about the same time as the German scholar Leibniz—developed the calculus.

The names of many great physicists and mathematicians are memorialized in engineering terminology, testimony to the critical role of science in the evolution of the engineering profession. Robert Hooke, physicist and mathematician, was a contemporary of Newton, and it was he who found that for solids, within the elastic limit, deformation is proportional to stress—we call it Hooke's Law, the essence of structural analysis. Another great scientist of the age was Robert Boyle, who set forth the basic concepts of modern chemistry and used a vacuum pump in discovering that the pressure and volume of a gas are inversely proportional to each other—the famous Boyle's Law studied by generations of mechanical engineering students. In the eighteenth century the Swiss mathematician Leonhard Euler did pioneering work in analytical mechanics and hydrodynamics. (We study Euler's formula for the horizontal deflection of columns under vertical load.) In 1727 he went to St. Petersburg at the invitation of Catherine I, succeeding to the post held by Daniel Bernoulli, he of the Bernoulli's principle that underlies the carburetor and the atomizer. As the eighteenth century drew to a close, new technological vistas were opened by the experiments of Galvani and Volta, and these two giants of science—along with their nineteenth century successors, Ohm, Ampère, Faraday, Joule, and Kelvin—have been immortalized in electrical engineering textbooks.

The historical relationship between science and engineering is replete with paradox. Engineering today is, in large measure, the application of science to technological problems. Yet for more than two hundred years, as the great scientific findings were set forth in books or revealed in lectures and proliferating science societies, most engineers resolutely looked the other way. Of course, the usefulness of the new theories was not always immediately apparent. But more than that, to the practicing technologist the attitudes of the artisan were so deeply ingrained, the reliance on traditionally tested methods so strong, that the new ways of thinking seemed risky and foolish as well as alien. When, in 1742, three mathematicians were commissioned to investigate cracks in the dome of St. Peter's, they were ridiculed by practical builders. "Michelangelo knew no mathematics," noted a contemporary critic, "yet he was able to build the dome."[1]

There was often good reason to be skeptical of the efforts of theoreticians. Frederick the Great, writing to Voltaire in 1778, spoke of a wheel that the famous Euler had mathematically calculated would lift water in his garden. It did not work. "Vanity of vanities!" exclaimed the prince. "Vanity of mathematics!"[2]

In 1805 a noted French architect announced that "in order to probe the solidity of buildings, the complicated calculations, bristling with figures and algebraic quantities, with their 'powers,' 'roots,' 'exponents' and 'coefficients' are by no means necessary."[3] In 1822 Thomas Tredgold, a celebrated British engineer who had worked his way up from journeyman carpenter, observed that "the stability of a building is inversely proportional to the science of the builder."[4] When, in 1858, W. J. M. Rankine of Scotland issued his famous and widely used *Manual of Applied Mechanics,* he sought to put an end to the deplorable "separation of theoretical and practical knowledge."[5] Yet as late as 1872 the author of a *Civil Engineer Pocketbook* stated that he would not refer to Rankine or other exponents of theory because they are "but little more than striking instances of how completely the most simple facts may be buried out of sight under heaps of mathematical rubbish."[6]

The wedding of science and engineering was eventually sanctified, though traces of the stormy courtship persist. There are those today who see technological advance in terms of scientific genius, as if engineering were simply the gross application of sublime theory. At the other extreme are the vociferous supporters of hands-on ingenuity. Some historians of technology take pleasure in crediting scientific advance mostly to craftsmen—to the lens-grinders and instrument-makers, to the people of the forge and workbench. These historians make the point that without instruments—from telescopes to atom-smashers—there would be precious little science. The level of technical accomplishment thus defines the range of scientific inquiry.[7] In addition to the crucial role of instruments there is also the tenacious work of the engineering experimenter—the tinkerer, if you will—that often turns up new facts and relationships long before they are scientifically understood. Aircraft, rocketry, turbines, and semi-conductors are just a few of the many fields in which engineering has led and science has followed.

On both sides the argument is waged largely through the use of imaginary caricatures. Most scientists have not been as absent-minded and metaphysical as they have been portrayed, nor technologists as muscled and unlearned. Leibniz, like many scientists of his time, was enormously interested in the development of machines, and in a letter dated 1707 proposed conserving waste steam, and even superfluous furnace heat, in what we today would call cogeneration.[8] On the other hand, the humble millwright of the eighteenth century, according to an apprentice of that trade, "was a fair arithmetician, knew something of geometry, levelling and mensuration, and in some cases possessed a very competent knowledge of practical mathematics. He could calculate the velocity, strength, and power of machines: could draw in plan and section. . . ."[9]

James Watt is often cited by proponents of both sides of the debate. He was, to be sure, curator of precision instruments at the University of Glasgow, tremendously gifted with his hands, and his work on the steam engine antedated much scientific theory on the behavior of gases. So one is tempted to agree with those who claim that the steam engine did more for science than science ever did for the steam engine. Yet it is equally true that Watt was friendly with many of the professors at the university and that he consulted with Dr. Joseph Black, who occupied the chair of anatomy and chemistry, on the phenomenon of latent heat. Like many artisans, inventors, and engineers, Watt was intensely interested in the scientific theories that related to his work. He and his business partner, Matthew Boulton, were members of the renowned Lunar Society and attended meetings along with scientists such as Erasmus Darwin and Joseph Priestley, as well as with technologists such as the potter Josiah Wedgwood.

From our present vantage point it seems only sensible to take a middle position on what makes for scientific progress, to credit the workshop as well as the laboratory, the hand as well as the mind, to value intuition and experience as well as theory and experiment. As for engineering, let us conclude once and for all that it is a blend of art, craft, and science.

The Renaissance, the Protestant Reformation, the Age of Exploration, the Age of Science, the Age of Reason—all these terms are historians' shorthand for vast and complex changes that occurred in Western civilization, changes that included, in great measure, a quickening interest in engineering progress.

Francis Bacon spoke eloquently of uniting science and technology and securing for both the support of an enlightened government. In Bacon's idealized society, portrayed in *New Atlantis* (1627), trained specialists worked harmoniously together in pursuit of scientific truth and "profitable inventions," all for the good of the commonwealth. This dream was partially realized in 1660 when the Royal Society was founded in London and subsequently granted a charter by Charles II. But meaningful funding did not accompany the Royal Charter, and the society's members were left pretty much to their own devices, bound only by the loose links of interest and curiosity.

The methodical pursuit of research and development required a more bureaucratically structured society than that of seventeenth-century Britain. The requisite wealth, order, and administrative sophistication were to be found across the Channel in the realm of the Sun King, Louis XIV of France. When the French government established the *Academie Royale des Sciences* in Paris in 1666, it buttressed its good intentions with financial support. Members were paid stipends and directed to apply scientific findings to technological problems. In the interest of national prosperity, French scientists were encouraged to think in terms of applications, and engineers were urged to make use of scientific concepts.

The French army provided a major source of engineering talent. In 1675, at the suggestion of Sébastien le Prestre de

Vauban, builder of fortresses and harbors (and later to be named a *Marechal de France*), the army created a special organization of military engineers—the *Corps des ingenieurs du Genie militaire.* It was at this time that the term *ingenieur,* which previously had connoted a craftsman-builder, began to take on the meaning of scientifically trained professional. These military engineers gradually gained responsibility for public works; when it came to bridges, roads, canals, and water supplies, it was often difficult to differentiate between communal benefit and military need. Eventually, in 1716, France established a civilian engineering corps, the *Corps des Ingenieurs des Ponts et Chaussees,* and starting in 1747, the government entrusted the education of these engineers to the newly created *Ecole des Ponts et Chaussees.* This was a significant event in the history of engineering. At the *Ecole,* mathematical theory was applied to matters that from time immemorial had been subject to rule-of-thumb convention.

Among the most prominent of the *Genie* officers was Charles Augustin de Coulomb, who addressed the problems of beam flexure, earth pressure on retaining walls, vaulting, and compression of masonry, and conducted research into the elasticity of wires and on electricity and magnetism. Coulomb became a member of the *Academie des Sciences* and was inspector of hydraulic works until the fall of the monarchy in 1792. After the Revolution, Napoleon appointed him a member of the *Institut National* and Inspector General of Public Education.

The tradition of government support for engineering survived the French Revolution. In fact, it was the revolutionary government, in 1794, that established the *Ecole Polytechnique.* That institution, soon to become world renowned, opened with four hundred students and boasted on its staff the noted mathematicians Lagrange and Laplace. A competitive examination, based chiefly on mathematics, dictated admissions. Out of this institution, and others of its kind, there evolved a technocratic elite that has ever since played a prominent role in French affairs.

By the end of the eighteenth century, French engineering—

particularly its bridges and roads—was the envy of all Europe. In chemistry, too, the government's support for research and development paid handsome dividends. The manufacture of gunpowder, the processing of sugar from beets, the manufacture of dyes and alkali—these were just a few of the fields in which France showed the way.

Yet it was in Britain, not France, that the Industrial Revolution burst forth. While French technological progress was tied to government support and bureaucracy, the British pattern was one of individual ingenuity and entrepreneurial initiative. Private family enterprises, most notably those of the Crowleys, the Crawshays, and the Darbys, controlled the fabrication of iron. Their kinship practices (granting responsibility and ownership to children, cousins, and in-laws), reminiscent of the Scottish clan system, provided effective administration for as many as five generations. These industrial clans invented, developed, and perfected many ironmaking techniques that are landmarks in the history of technology. A French observer, in 1786, complained of his own nation's backwardness in this "most essential of trades" and grudgingly spoke of the English "skill in working iron and the great advantage it gives them as regards the motion, lastingness, and accuracy of machinery."[10]

The other vital component of the Industrial Revolution was, of course, the steam engine. In 1698, one Thomas Savery of Devon patented a water pumping device consisting of a cylinder that was alternately filled with steam and then chilled (by pouring water on the outer surface), creating a vacuum that sucked water through a pipe from below. Then a new blast of steam forced the water up through a pipe above. In 1712, a more effective engine (powered by a piston pushed by atmospheric pressure into an evacuated cylinder) was produced by Thomas Newcomen, an ironmonger also from Devon. For more than half a century, Newcomen engines served to drain British mines, albeit with woeful inefficiency. Starting in 1768, they were gradually replaced by James Watt's radically improved design. Within another twenty years the steam engine had been applied to newly developed spinning machinery, marking the beginning of the factory age.

The growth of British invention as described in history books sometimes seems like a succession of spontaneous personal inspirations. Actually, progress came in hard-won increments, entailing many experiments by many teams of technicians. And though the British Crown was not nearly as supportive as was the government in France, it is not true, as some historians suggest, that it played no role at all. The Royal Ordinance Department maintained at Vauxhall a complex of foundries and shops manned by a cadre of trained craftsmen and engineers, and many of the mechanical advances leading to the steam engine were accomplished there. Equally important were constructive government policies granting patents and monopolies and fostering finance. The Bank of England, founded in 1694, antedated its French counterpart by more than a century.

Few of the millwrights, mechanics, ironworkers, and just plain tinkerers of eighteenth-century Britain thought of themselves as engineers. The concept of professionalism evolved earliest, as it did in France, among the constructors of public works. John Smeaton, builder of bridges, harbors, and lighthouses, began to call himself a "civil engineer" around 1750. The term was intended to rid his profession of the disagreeable association with military affairs. Although he made his reputation in great construction projects—most notably the Eddystone Lighthouse—Smeaton also worked on improving waterwheels, steam engines, boring machines, and other mechanical devices. Truly the complete engineer, this remarkable man was made a Fellow of the Royal Society. He also evinced a strong sense of professional pride, and on March 15, 1771, he met with several of his colleagues at the Kings Head Tavern in Holborn to form the Society of Civil Engineers (soon to be known as the Smeatonian Society). Social clubs were very much the vogue in Britain at that time. Many groups gathered to exchange scientific and technical information, and many societies were formed, some by scientists, engineers, or manufacturers, others by what today we would call Chamber of Commerce "boosters." It is known that Smeaton attended at least one meeting of the Lunar Society and several of that organization's members were also Smeatonians.

British engineers manifested a propensity to splinter into camps professing subtle distinctions of approach, a characteristic of the profession to this day. The Royal Institution was founded in 1799 to facilitate "the general introduction of useful mechanical inventions and improvements." The Society of Millwrights appeared in 1805. In 1818 the Institution of Civil Engineers was formed, born of a revolt against the excessively scientific approach of the Smeatonians. The Institution of Mechanical Engineers, so the story goes, was founded in 1846 because of an alleged affront by the civil engineers to George Stephenson, a noted builder of locomotives.

While British engineers were developing a sense of professional identity, certain social visionaries began to advocate scientific training for working artisans. By 1800 so-called "steam-intellect societies" were being formed for this purpose and by 1850 there were 700 mechanics' institutes in Britain catering to 120,000 students. There is little evidence that this education did much to improve the mechanics' lot, either professionally or socially. Nevertheless, the nation eventually benefitted when several of the institutes developed into notable engineering schools. Less auspiciously, the movement strengthened the popular perception that technological pursuits were most properly the province of the lower classes.

The identification of engineering with lack of social status was reinforced by the religious prejudices of the time. In eighteenth-century England religious non-conformists (those individuals who were not members of the Church of England) were barred by the universities and the learned professions and even ostracized in the older towns and cities. As a consequence, many independent-minded artisans set up business in the midlands and to the north, away from the centers of orthodoxy, and pursued careers in invention and manufacture. As noted by several historians of technology, the refusal to conform requires great determination, individuality, and strength of mind, the very characteristics that are required for success in technological enterprise. French science and engineering also were affected by policies of intolerance. After the revocation of the Edict of Nantes in 1685, many skilled Protestant artisans and scientists fled, creating a serious "brain drain" almost three hundred years before that phrase was coined. The American colonies warmly welcomed skilled workers and professionals fleeing European oppression.

History reveals among engineers a free-thinking, pragmatic streak of independence. When engineers are criticized for their willingness to cooperate with government authorities, it should also be remembered that they have often chosen to go where life is free and congenial, and to take their skills with them.

In the contrast between France and Britain we see the seeds of dichotomies and debates that characterize engineering today. As if to prove that there is no one true road to technological superiority, France and England waged fierce battles—both military and mercantile—without reaching a decisive conclusion. In the short term England won. Napoleon suffered defeat at Waterloo in 1815 and the glory years of the British Empire followed. Much credit for this must go to British engineering and entrepreneurial genius—as well as to the nation's unique geography and its natural resources, particularly coal. But by the time of the Great Crystal Palace exhibit of 1851, knowledgeable British observers were concerned by the great technological advances that nations of the Continent were making; at the International Exhibition of 1867, held in Paris, their concern turned to alarm. British products received a bare dozen awards. It was clear that the empiric skills of the British engineers were not adequate to the new era. Chemicals, steel, electricity—these industries required an abundance of scientifically trained engineers. While practically all British engineers continued to be trained by apprenticeship—the only university chairs of engineering were at Glasgow, Edinburgh, and University College in London—the concept of the French *polytechnique* was taking hold throughout a technologically resurgent Europe.

In Germany, especially, the changes were astonishing. That nation had, by the late nineteenth century, transformed itself from an agricultural backwater into a mighty power boasting four outstanding research-oriented institutions and a dozen engineering colleges, supplemented by mining academies, technical high schools, and military schools. The collaboration between German industrialists, engineers, educators, bankers, and government officials was unparalleled in human affairs. When, in 1871, Germany defeated France and overthrew Napoleon III, Disraeli remarked, "This represents the German Revolution, a greater political event than the French Revolution of the last century."[11] Unfortunately—and not for the first time in history—technological accomplishment was a prelude to military aggression.

Britain—fountainhead of the Industrial Revolution, home of Telford and Rennie, giants of bridge building; Brunel, conceiver of steamships; Stephenson, father of locomotives; the native land of Smeaton, Watt, and McAdam—had forfeited forever its preeminence. The very independence of its institutions, including the apprenticeship practices of the engineers and their societies, had impeded the requisite cooperative national effort. Not until 1889 did Parliament provide substantial funding for technical education, mainly through grants to city universities. Cambridge and Oxford followed reluctantly. A program in "mechanical science" was first introduced at Cambridge in 1890; a chair of "engineering science" was established at Oxford in 1909. This foot-dragging is hardly surprising when one considers that as late as 1850—a century and a half after Isaac Newton became President of the Royal Society—the "pure" physical sciences had scarcely gained a toehold in those hallowed halls.

The economic decline of Britain demonstrates that engineering superiority can be lost as quickly as it is gained, and that in addition to energy, inventiveness, and good intentions, technological vitality cannot be maintained without promotion of science, support for engineering education, and enlightened cooperation between government, industry, and academia.

On the North American continent, technology evolved out of the needs of frontier living. Tools and techniques were brought from Europe, but the relative isolation and the immediate problems of pioneering early encouraged the flowering of "Yankee ingenuity." Professional engineering was virtually non-existent. As George Washington told John Randolph, anyone wishing to dig a canal or build bridges "must invite a proper person from Europe."[12] One of the earliest of such "proper persons" was William Weston, an apprentice of the English canal builder James Brindley. Between 1793 and 1801 Weston consulted on the building of locks on the Mohawk River at Rome, New York; drew a plan for a water supply system for New York City; made recommendations for the reconstruction of the Lancaster Turnpike; and designed piers for the Market Street Bridge in Philadelphia. Another prominent English immigrant was Benjamin Latrobe, a student of John Smeaton, who worked on the Philadelphia waterworks, a Susquehanna navigation survey, and various canals, as well as acting as architect on many large domed buildings. Robert Fulton was born in Pennsylvania in 1765, but studied and worked in Europe, returning to the United States at the age of forty to design his steamboat.

When the Erie Canal was begun in 1817, there were perhaps thirty men in the entire nation who considered themselves qualified as engineers, and a good number of these were little more than surveyors or builders. The commissioners entrusted with building the canal, unable to recruit a true professional either domestically or from abroad, decided that the construction of canals was "an art within the complete attainment of ordinary capacities"[13] and designated two native-born surveyors, Benjamin Wright and James Geddes, as engineers in charge. The work proceeded, willy-nilly, and during the next eight years a pattern was established for recruiting assistant engineers from within the ranks of the various surveying parties. The New York canal system became, in effect, an on-the-job training ground for engineers, a variation of the English apprentice system. By 1825, when the Erie and Champlain Canals were completed, thirty more engineers had thus entered the profession, doubling the nation's supply.

Another engineering resource, as in France, was the army. The United States Military Academy was founded in 1802 as a school for engineer officers. (An act in 1812 broadened its mandate to train officers for all units of the army.) The Academy's first superintendent, Sylvanus Thayer, appointed to the post in 1817, traveled to Europe to study military engineering and education. He introduced a curriculum based upon the polytechnic concept, and soon the academy was turning out engineers with excellent academic credentials. One responsibility of the Army Corps of Engineers was to perform surveys of harbors and rivers, and occasionally it was authorized to do surveys for roads and canals considered to be of national importance. But this work was limited due to congressional concern about using the army for civilian—and often privately sponsored—projects. Before long, however, West Point graduates, after retiring from the army, began working as civilian engineers, their number increasing from fifteen in 1830 to more than one hundred in 1838.

These ex-military engineers were useful, but the rapidly developing nation clearly needed civilian engineering schools. The example of Europe was not lost on intelligent observers, although as in England, the traditional universities were hostile to the concept. So were most of the practicing engineers who themselves had been trained in the field or in the shop. The initiative was left to civic-minded individuals like Stephen Van Rensselaer of Albany, land owner, capitalist, and leader in public affairs, who in 1823 took steps to establish a school at Troy "for the purpose of instructing persons who may choose to apply themselves in the application of science to the common purposes of life."[14] Under the direction of Amos Eaton—lawyer, engineer, and scientist—the Rensselaer School, as it was first called, developed in the course of twelve years into a professional school of engineering, the degree of civil engineer being first bestowed on four graduates in 1835. In 1849 the school was reorganized by its new director, B. Franklin Greene—who, like Thayer at West Point, had traveled to Europe and been impressed by the French model—and the name Rensselaer Polytechnic Institute was adopted. By 1860, 318 RPI alumni were practicing professionals, along with almost 200 civilian graduates of the military academy.

Slowly, the example set by Rensselaer was followed by other institutions. Union College introduced civil engineering courses in 1845, and the University of Michigan in 1847. In the same year both Harvard and Yale took steps to create schools of applied science, halting steps though they were. The Lawrence School at Harvard—endowed by a manufacturer of woolen goods—graduated only forty-nine men before the Civil War, and this, according to an engineering educator, "in the face of an unconcealed disdain on the part of the regular faculty."[15] This attitude helped prompt the establishment of the Massachusetts Institute of Technology, which obtained land in a charter from the Massachusetts State Legislature in 1861 and opened its doors four years later.

The big breakthrough in American engineering education occurred with the passage of the Morrill Act (popularly known as the "land grants" act) in 1862. This legislation, passed in the midst of the Civil War, granted Federal aid to the states for founding colleges of agriculture and mechanic arts. Ironically, the bill had been pending in Congress for five years, and if not for the absence of the Southern states, its passage probably would have been further delayed. This legislation, coupled with the economic expansion that followed the war—particularly the boom in railroad building—inspired a period of spectacular growth in engineering education. Not only were new "land grant" colleges founded, but many of the older universities added departments of engineering. The number of schools teaching engineering increased to seventeen by 1870, to forty-one by 1871, to seventy by 1872, and to eighty-five by 1880.

Still, among practicing engineers, the concept of academic learning was slow to take hold. In 1870 only one in nine American engineers was a college graduate, and as late as 1916, incredible as it may sound, the figure was only about one in two. In many quarters, the school-taught engineers were regarded with suspicion by the old-timers.

Another peculiar aspect of the growth in engineering academia is that it coincided with a decline in the social status of the profession, indeed in some ways contributed to that decline. Although RPI developed along the lines of the elite French polytechnics, its original statement of purpose spoke of "instructing the sons and daughters of farmers and mechanics." The founding legislation of the Morrill Act mentioned "education of the industrial classes in their several pursuits and professions in life." And the stress on vocationalism that prevailed in many engineering school curricula added to the general impression of engineering as a respectable but somewhat lowly occupation.

Up to the mid-nineteenth century, engineering had attracted "many sons of leading American families," sons of physicians, attorneys, judges, clergymen, academics, and wealthy businessmen.[16] The engineers themselves were often intellectuals and social visionaries, living refutations of Plato's stereotype. John Roebling, designer of the Brooklyn Bridge, and his son Washington Roebling, who completed the project, were multi-talented scholars whose erudition helped to make them widely respected public figures. This patrician quality was not shared by the young men who started to crowd into the land grant colleges.

Ten years before the elder Roebling's death in Brooklyn, Britain's great railway engineer, Robert Stephenson, died and, according to L.T.C. Rolt, "the whole nation mourned him."

> By special permission of the Queen, his funeral cortege passed through Hyde Park on its way to Westminster Abbey where his body was buried beside Thomas Telford, and the whole route was lined by silent crowds. In his home county, all shipping lay silent on Tyne, Wear and Tees, all work ceased in the towns and flags flew at half-mast. . . . It was as though a king had died, but where was his successor?
>
> Never again would a British engineer command so much esteem and affection; never again would the profession stand so high.[17]

Stephenson in Britain, the Roeblings in the United States, Eiffel of the famed tower in France—a few such men of mythic reputation flashed across the pages of history. But, as civil and mechanical engineering became increasingly mathematical, and as chemical and electrical engineering burgeoned into new disciplines, the epic figure of the "chief engineer" faded from the scene. No single person could master more than a small portion of the exploding fund of knowledge. Engineering, by its very success, transformed itself into a group enterprise, without individual heroes.

A number of supremely talented individuals continued to make unique contributions. Edison, Westinghouse, Bell, Eastman, Diesel, Marconi, the Wright brothers—most of them inventors rather than engineers in the usual sense—became famous, indeed legendary. But after them a curtain of anonymity settled over the profession. Herbert Hoover ran for the presidency as "The Great Engineer," but that single event shed little light on other engineers and in the end very little glory on engineering. There is today a Hoover Dam, but few people can name its designer—or the designers of the Grand Coulee Dam or the Verrazano Narrows Bridge, the creators of radar or the transistor, the developers of air conditioning or the space shuttle. The mighty corporations are famous, as are the laboratories and universities where the frontiers of engineering are expanding at an accelerating rate. Mention IBM, RCA, GE or Exxon, speak of M.I.T., Cal Tech, Bell Labs or Los Alamos, and a light of recognition flares. But the leaders of these institutions—even the Nobel Prize winners among them—are scarcely known beyond the circle of their colleagues.

For a while, even as individual engineers sank into anonymity, the profession as a whole retained an aura of heroic distinction. During the first half of the twentieth century, technology held the promise of wealth and ease for all, and engineers were the clever chaps who made technology jump through hoops.

After World War II, however, as we have already noted, public perception of the engineer began to change. Nuclear weapons created a widespread malaise. The 1960s brought us Vietnam, napalm, and Agent Orange; the 1970s featured environmentalism, asbestos, and toxic wastes. The changing public mood was influenced by the critiques of such disenchanted observers as Rachel Carson, Ralph Nader, and Barry Commoner. There was applause, to be sure, for moon shots, computers, and satellite-transmitted television, but the image of the engineer became badly tarnished.

This decline in reputation was compounded by a loss of professional independence. As the corporate structure of American industry grew, more and more engineers became employees of giant organizations. An ever smaller proportion carried on as autonomous consultants, and hardly any continued to play the time-honored role of independent inventor. Still, the adventurous, visionary virtues survived and have seen something of a renaissance in the entrepreneurial Wild West of Silicon Valley. But the vast majority of today's engineers find themselves in the midst of multitudes, worrying about workers' rights or, at best, struggling up the slippery ladders of middle management. Given the nature of modern technology and the dynamics of corporate growth, it seems that events could hardly have been otherwise. Some commentators allege that engineers "sold out" to the corporations, becoming an exploited underclass. This view, I believe, mistakenly describes a spontaneous, natural process in terms of capitalist villainy. Nevertheless, the melancholy truth remains: as engineering grew and flourished, it forfeited autonomy and glory.

This historical trend is especially poignant when one considers the attempts that were made to fashion a noble and independent profession. The loftiest aspirations were expressed by the founders of the various engineering societies, starting with The American Society of Civil Engineers in 1852. The American Institute of Mining Engineers was established in 1872, the American Society of Mechanical Engineers in 1883, the American Institute of Electrical Engineers in 1884, and the American Institute of Chemical Engineers in 1908. Within each of these organizations engineers endeavored to elevate scientific and professional standards, emphasize ethical and social concerns, and garner a portion of public esteem.

Battles were fought between those who sought to insulate engineering from business and those who believed that the profession's proper role was to become part of industry, albeit in positions of professional leadership. Both views had strong and unyielding proponents who filled the professional journals with impassioned articles and letters. Looking back, one is inclined to conclude that both groups had much reason to be satisfied with their lot. Independent consulting firms prospered, while in corporations engineers advanced to positions of responsibility. By the 1920s, two-thirds of the graduates of American engineering schools were becoming managers within fifteen years of graduation. A survey of the profession reported "a healthy progression through technical work toward the responsibilities of management."[18] It was not foreseen that industry's appetite for engineers would become increasingly voracious, and that inevitably large numbers of engineers would become stuck in the lower ranks.

Engineers walked willingly into the corporate jaws, entranced not only by attractive salaries and technical challenge, but also by visions of bringing "efficiency" to factory and office. In the current era of prestigeous business schools we tend to forget the brief but intense passion with which engineers claimed management as their own. Around the beginning of this century, Frederick W. Taylor, a mechanical engineer from a well-to-do Philadelphia family, studied the work habits of factory personnel and proposed methods of maximizing production. Known first as the "Taylor System," these ideas were disseminated broadly as "human engineering," "industrial engineering," and "scientific management." For a few years this new specialty—remembered today chiefly for its time and motion studies—achieved great popularity. But labor groups resisted what they considered to be an exploitive approach, and many engineers regarded the movement as a perversion of engineering methodology. By 1916 Taylor's methods were banned in government-funded operations and the vogue quickly dissipated. Nevertheless, engineers continued to play a role as planners and overseers of technical work. A course called Engineering Administration was established at M.I.T. in 1913. In 1932 this specialty became a separate department in the School of Engineering. In 1952, however, the department was transmuted into an independent School of Industrial Management. Today the growth of business administration as a separate discipline has served to draw a line between engineering and management, although the distinction can never be absolute.

Out of this interest in scientific management grew the idea of utilizing engineering methods to solve social problems. Around the time of World War I, a few so-called "progressive" engineers met to discuss the possibilities of applying Taylor's principles to the reorganization of society, but the group failed to make much of an impact on professional or political affairs.

According to census reports, there were just 512 engineers in the United States in 1850. By 1880 the number had risen to 8,261. After this the census figures become deceptive because at various times they included locomotive engineers, mechanics, surveyors, and electricians. Other records, however, indicate that in 1900 there were approximately 45,000 engineers, and in 1930, 230,000. Even at that point, according to one study, "the profession remained tiny and elite."[19]

Today the figure has risen to well over a million and a half, with more than a hundred thousand college freshmen entering engineering programs each year. The unique role of yesteryear's professional could scarcely be expected to survive under the crushing weight of such numbers.

In many ways the American engineering profession today is riding high. And yet . . . When we think of the great numbers, the increased specialization, the faceless anonymity, and the submersion of engineering into corporate culture, we may well begin to yearn for days gone by.

Whenever I find myself wishing wistfully for the engineering grandeur of the past, I remind myself of the time my wife and I were vacationing in Italy at the height of the holiday season. We joined a tour of the Vatican and as we arrived at the Sistine Chapel found ourselves jostled by an oppressive crowd. We, and others in our group, could not help expressing disappointment and frustration. But our guide, an energetic little woman who carried on high a rolled newspaper to make sure we did not lose sight of her, insisted that we grasp the magic of the moment. "Look up," she said. "Ignore the crush about you and concentrate on the magnificent paintings that you see. Michelangelo created them for you, and they could not be any more divine if you stood here alone with the artist by your side." Her admonition had its intended effect. The great paintings *are* there, and the feelings they elicit transcend time and circumstance. I look upon engineering and its history in somewhat the same way. The profession endures—in all its essentials—for those who would find in it pride and inspiration.

Where do we come from and where are we heading? There is no single truth that embodies the practice of engineering, no patron saint, no motto or simple credo. There is no unique methodology that has been distilled from millenia of technological effort. Contradictions abound. Indeed, the essence of engineering lies in its need and willingness to embrace opposites. Empiricism and theory, craftsmanship and science, workshop and laboratory, apprenticeship and formal schooling, private initiative and government venture, commerce and independent professionalism, military necessity and civic benefit—all of these and more have their place. And running through the profession's richly textured history are those perplexing questions of class and prestige, the elite and the plebian.

Engineering stands in the turbulent center of democratic life, thriving on variety, vital in the midst of paradox. If the past is indeed prologue to the future—if simple craft can evolve, as it has, into engineering science—there is ample reason to hope for the coming of ever more accomplished engineers, indeed for the coming of another engineering renaissance the nature of which we can as yet only dimly envision.

5

The Engineering View

What *is* an engineer? The National Research Council's Committee on the Education and Utilization of the Engineer recently framed a definition:

> *Engineer.* A person having at least one of the following qualifications:
> a. College/university B.S. or advanced degree in an accredited engineering program.
> b. Membership in a recognized engineering society at a professional level.
> c. Registered or licensed as an engineer by a governmental agency.
> d. Current or recent employment in a job classification requiring engineering work at a professional level.[1]

Very well, then, what is *engineering?* Again the committee provides an answer:

> Business, government, academic, or individual efforts in which knowledge of mathematical and/or natural sciences is employed in research, development, design, manufacturing, systems engineering, or technical operations with the objective of creating and/or delivering systems, products, processes,

and/or services of a technical nature and content intended for use.[2]

Quite a mouthful, and not really an improvement on Thomas Tredgold's classic definition of 1828: "Engineering is the art of directing the great sources of power in nature for the use and convenience of man."[3]

Engineers, as viewed by the public and as confirmed by self-image, constitute a single profession. Yet the roughly 1.6 million American engineers form a strikingly diverse group. First of all, they differ by specialty in the following way:

Discipline	Percent Practitioners[4]	Percent Degrees 1985[5]
Electrical/Electronic	18.9	28.5
Mechanical	16.9	22.0
Civil	13.5	12.2
Industrial	9.7	5.6
Chemical	5.4	9.3
Aero/Astro	3.4	3.4
Other, including mining, computer, manufacturing, petroleum, marine, agricultural, nuclear, bio-engineering, engineering mechanics, environmental, ceramic, metallurgical, and materials	32.2	19.0
	100.0	100.0

Contrasts appear also when the profession is categorized by type of activity:

Activity	Percent[6]
Research	4.7
Design and Development	27.9

Activity	Percent
R & D management	8.7
Other management	19.3
Teaching	2.1
Production/Inspection	16.6
Other, including consulting, reporting, statistical work, and computing	20.7
	100.0

And even these categories do not begin to hint at the diverse social roles that engineers play. Consultants, researchers, and academics; entrepreneurs and corporate executives; public works officials and other civil servants; a wide variety of quasi-professional employees—the spectrum is broad and getting broader as computer engineering attracts a new breed of bearded blue-jeaned mavericks. As for politics, there are right-wing engineers and left-wing, hawks and doves, pro-growth and anti-growth, pro-nuke and anti-nuke. Some engineers are ardent environmentalists; others are hostile to environmentalism. Most engineers, like most citizens, hold many different opinions, some of them inconsistent and subject to frequent change. Needless to say, there are a variety of religious beliefs and ethnic affiliations. Engineering used to be something of an all-male association, but this is no longer the case.

Even before the profession exploded into a myriad parts, engineers had difficulty in getting together—or rather staying together—socially and institutionally. One might think that an engineer would be the quintessential organization person, but the truth is otherwise (a fact touched on in the previous chapter in my discussion of the first British engineering societies). In the United States the splintering phenomenon started early and continues to this day. There is no powerful, central, national engineering organization comparable to the American Medical Association or the American Bar Association, and there are good reasons to believe that there may never be one.

Since the founding of the American Society of Civil Engineers (ASCE) in 1852, the main organizing thrust of the engineering profession has been along the lines of technical specialties. All efforts at profession-wide federation have ended in failure. The latest attempt—the American Association of Engineering Societies (AAES), instituted in 1979—almost succumbed to intramural feuding, and exists today as a coordinator of loosely affiliated member societies. "When we started," says a member of the AAES board of governors, "we thought we would hire someone who could speak for the entire engineering community. But that was very difficult because the engineering community didn't have any discrete message."[7]

Beyond the hard facts of diversity, however, and overriding the details of definition, is there not for engineers a shared way of approaching the world, a common outlook that one might call "the engineering view?" I suggest that there is.

In seeking the essence of the engineering view it seems appropriate to begin with the scientific view. All contemporary engineers enter their profession by passing through the portals of science. In order to be admitted to an accredited engineering school, a young person must have studied science and shown an aptitude for it. To graduate from an accredited engineering school he or she must study a lot more science—and, of course, the mathematics that is an essential element of science. Indeed, the almost exclusive criterion for entering the engineering profession is the ability to "do" science and to do it well.

"Doing" science implies a belief in science, and I think it is fair to say that this belief lies at the heart of engineering. The engineer does not believe in black magic, voodoo, or rain dances. The engineer believes in scientific truth, that is, truth that can be verified by experiment.

The search for scientific truth requires that we disregard, so far as possible, our personal value systems; yet, paradoxically, this approach creates its own values. As Jacob Bronowski has written: "Independence and originality, dissent and freedom and tolerance; such are the first needs of science; and these are the values which, of itself, it demands and forms."[8] In the same vein, Bertrand Russell asserted that "those who forget good and evil and seek only to know the facts are more likely to achieve good than those who view the world through the distorting medium of their own desires."[9]

So this is the beginning of the engineering view: a commitment to scientific truth and to the values that the search for this truth entails.

Occasionally our critics view this commitment as evidence of a lack of "soul." When one deals in hard facts it is easy for an observer to conclude that one has a hard heart. Using an even uglier metaphor, Theodore Roszak has said that scientists and technologists look at the world with "a dead man's eyes."[10] I find such criticism to be particularly irritating. There is a place in the world for poetry and sermons and for visions of "what might be" rather than "what is." But there is also a need for facts and plain-speaking. As engineers we are pledged not to engage in merely wishful thinking. We are not the grasshopper; we are the ant who knows that winter is coming. We are the grumpy little pig who builds his house out of brick while his friends play and sing "Who's afraid of the big bad wolf?" We do this because we know what the big bad wolf can do when he huffs and puffs. This does not mean that, as individuals, we cannot love poetry and approve of sermons—or even preach sermons. But, because we are committed to scientific truth, we believe that poetry and sermons alone are not an adequate foundation on which to build human society. The evidence convinces us that God helps those who help themselves. It does not follow that we are selfish. The evidence also shows that God helps those who work together cooperatively and provide for one another.

There is a fringe benefit that comes along with our familiarity with science—and with the technological applications of science —and this is that it helps make us feel at home in the world. To the extent that the forces of nature have been comprehended, and the structure of the universe revealed, we share in the understanding and this gives us some measure of contentment. This comfort—this inner peace, if you will—is a basic ingredient of the engineering view.

This does not mean that engineers are, or have any reason to be, smug. We are humble before the unknown and stand in awe of the unknowable. But we do not feel alienated, as some people say they do, by the scientific advances of our age. And our message to our non-technical fellows is that to a certain extent this understanding—and the peace that goes with it—is available to all. Anyone who is willing to explore, however superficially, the findings of science, can share in this feeling of at-homeness in the world.

The study of science, if it is to lead to professional competence, must entail hard work, and the engineering view accepts this. Engineers believe in hard work. We demand it of ourselves and require it of those who would join our ranks. We think that hard work is somehow treasured in the scheme of things. Knowledge and understanding are precious objectives worth striving for. The quest for excellence is a virtuous enterprise that needs no rationalizing.

Although we are committed to scientific truth, there comes a point where this truth is not enough, where the application of truth to human objectives comes into play. Once we start to think in terms of utility, we must leave the placid environment of the laboratory, take off our white coats, and roll up our sleeves. We are no longer considering theoretical forces and ideal substances. We are now obliged to work with materials that are real, impure, and sometimes unpredictable. Our aim is no longer to discern absolute truth, but rather to create a product that will perform a function. And suddenly we find ourselves under constraints of time and money. To a practicing engineer the search for perfection becomes self-defeating. In *The Soul of a New Machine,* Tracy Kidder recounts how the engineers designing a computer must forego the luxury of constant refinement of their work. Instead of trying to make the perfect computer, they strive to make a good machine that they can get "out the door." They live for the moment when they can merely say, "Okay, it's right. Ship it."

If engineers are not afforded the luxury of seeking perfection, it follows that they must be willing to risk failure. Indeed, the willingness to make decisions knowing that something may go wrong is one of the most challenging aspects of the engineering experience. Engineers do not *want* to take chances. Prudence is implicit in our every undertaking. But as human societies seek to develop high civilizations, they inevitably strive to span ever wider chasms, build taller buildings and broader domes, create more sophisticated machines, and tap ever more elusive sources of energy. Citizens say to their engineers, "Do these marvelous deeds, but be careful." Engineers accept this charge knowing full well that every time they undertake a new design they may be defeated by unknown causes. We talk lightly of gremlins and Murphy's Law, but this banter is a manifestation of our underlying anxiety.

Even the most cautious engineer recognizes that risk is inherent in what he or she does. Over the long haul the improbable becomes the inevitable, and accidents *will* happen. The unanticipated will occur. Not that it is difficult to design redundant safety features into a product—it is merely expensive. It would be a lot easier for engineers if their fellow citizens would clearly stipulate that safety should be the paramount concern, whatever the cost. But the people do not say this. They want automobiles that are affordable and attractive; they want airplanes that are light enough to conserve fuel, and power plants that will turn out cheap electricity. They appear willing to pay for relatively foolproof backup systems for space vehicles, but precious little else. In other words, people are willing to take risks, but, naturally, do not want to pay the penalty for taking those risks. In such a world it requires a certain amount of moral toughness, something that verges on bravery, to say, all right, I will take the responsibility, and if the worst should happen, I will accept the blame.

Civil and mechanical engineers are particularly vulnerable since they have to cope with the devilish phenomenon of metal fatigue. There is no completely reliable way to predict how metals will behave when subject to vibrations and erratic stresses in changeable climate and over long periods of time. Sometimes the only way to gain knowledge is by experiencing failure. The British Institution of Mechanical Engineers published a fascinating volume entitled *Engineering Progress Through Trouble.* Reading this and similar works is a humbling and frightening—but in the end exhilarating—experience. To be willing to learn through failure—failure that cannot be hidden—requires tenacity and courage.

It has been said that doctors bury their mistakes and architects plant ivy. Most experts have a way of avoiding blame by claiming that their ideas were not given a fair trial. Economists are famous for this ploy. Politicians, as is well known, never make a mistake. Engineers have no such easy evasions. Well, so be it. Somebody has to step forward to do what needs doing. We can't all sit around being critics, supervisors, and second-guessers. Thus a principal feature of the engineering view becomes the willingness to accept responsibility.

Those citizens who take on responsibility make an implicit promise to their fellows to be *dependable.* You can count on us, we say to our neighbors. We will do our very best not to let you down.

Let me be clear about what I mean by this. I do not believe that it is up to the engineering profession to decide what is good for society, to decide, for example, whether we should favor mass transit or individual automobiles, allow drilling for oil off our coasts, authorize the use of public lands for mining, or determine how much of our national product should be devoted to armaments. In other words, I do not believe in technocracy. (I will return to this topic in the chapters ahead.) What I do mean is that in the pursuit of goals established by the entire community, engineers should be dependable. This commitment to dependability is something that we share with all the professions. It is, in fact, the essence of professionalism and it underlies our system of accreditation and licensing. I once knew an elementary school teacher who said that whenever she wanted to make sure that a specific task would be accomplished—erasing the blackboards after class, say, or putting the blocks away neatly, or bringing in the juice and cookies—she would call upon, if possible, a red-headed boy. She didn't know why, but in her experience red-headed boys were always dependable. In my experience engineers are singularly dependable. The resolve to be dependable is another essential element of the engineering view.

Obviously, engineering is dependent upon a society that functions effectively. Most engineers now work in large organizations and participate in mammoth undertakings. Thus, they learn to value cooperation and recognize that their work requires an established social structure. This is one way in which the engineering view differs markedly from the view of the artist. A poet can be a rebel—indeed, one of the main functions of a poet is to be a rebel, to startle and to shock. But engineering and rebellion do not go together. If the power plants are to operate and the factories to produce, law and order must prevail. In this sense engineering is conservative. Engineering depends upon social stability and contributes to it. Engineers are not likely to be anarchists or nihilists; that would be almost a contradiction in terms. There can't be any engineering in a chaotic world. But just because we believe in order does not mean that we, as a group, are committed to any particular type of order.

It has been said that because so many engineers are employed by American industry the profession has become a tool of American capitalism. The people who make this charge conveniently disregard the many thousands of engineers who work for federal, state, and local government, overseeing the nation's vast public works infrastructure and participating in essential regulatory activities. Increasingly, engineers are playing a role in public interest organizations, and of course many engineers are academics and private consultants. Even though engineers may agree that some organization is essential, it is clear that engineers have a variety of ideas about how society should be organized.

Most engineers tend to be pragmatists rather than ideologues. Ideology, after all, does not cure diseases or provide food and shelter. As pragmatists we are constantly re-evaluating our ideas about the social structure, and as we observe how things are done in Japan, for example, our interest in the possibility of change is surely piqued. But we must keep reminding ourselves that though we are dedicated to the concept of order, political ideology has no place in the engineering view. Individual engineers may be—should be—political, but they have no right to expect that their colleagues, *as engineers,* will share their partisan commitments.

Although engineering requires social stability and engineers are rarely found amoung the ranks of revolutionaries, technological progress has, in fact, contributed to the growth of democracy. Democratic ideals were expressed many centuries ago, but only modern engineering has made it possible for these ideals to be brought closer to fulfillment. It may not sound impressive to say that the goal of our society is "a car in every garage," but this is merely shorthand for saying that one person is as good as another and that opportunity should be available to all. The accumulation of material wealth has made it possible to think in this vein. Printing, radio, and television have made it possible and perhaps inevitable for the democratic ideal to spread and eventually prevail. Education and culture—not just for the privileged few but for the many—have followed in the wake of technological development. I do not mean to imply that engineers sally forth each day to save democracy. But in subtle yet powerful ways their work has served this cause. People have expressed concern lest high technology bring about despotism by making it possible for the few to dominate the many. But the evidence indicates that this is not happening and will not happen. The flame of freedom, once ignited, is almost impossible to extinguish. At least this is so where technology provides people with the basic necessities of life. Poverty and hunger serve the ends of despotism, and engineering fights the good fight against these scourges.

Engineering serves the cause of democracy in another way by being in itself—in its institutions and its personality—democratic. As noted above, Jacob Bronowski has said that science requires "independence and originality, dissent and freedom and tolerance"; engineering shares these needs. And in the United States most engineers have come from the middle and lower middle classes. Engineers are not snobs. We may sometimes be dull, but we are hardly ever snobbish.

Speaking of being dull, there can be no denying that the engineering view is essentially serious. Engineering work involves logic and precision. Unfortunately, this can lead to coldness and austerity. Engineers have a reputation for being humorless, and I fear this reputation is not entirely unfounded. There is no sense pretending that engineering is fun and games (although it often *is* fun), or that engineers are by nature jolly. We can certainly hope that our seriousness will be of good quality, that we will be earnest without being morose, sober without becoming glum. Maybe we can even try to become slightly more lighthearted.

Although engineering is serious and methodical, it contains elements of spontaneity. Engineering is an art as well as a science, and good engineering depends upon leaps of imagination as well as painstaking care. Creativity and ingenuity, the playfulness of original ideas—these are also a part of the engineering view.

Since engineering is creative—persistently and energetically creative—it has quite naturally become identified with the concept of change. In the American mythos, the quality of life can be constantly improved, an outlook that both stems from, and contributes to, the engineering view. However, we are not as naïvely optimistic in this regard as we used to be. It is now widely recognized that not all change is for the better and that change merely for the sake of change is foolishness. Human beings, we now recognize, are resistant to extensive alterations in their patterns of life. Just a few years ago a typical vision of the future pictured puny people seated in front of glowing TV screens and taking their meals in the form of pills. What actually happened in this era of high technology was an astonishing growth in jogging, body-building, and exercising of all sorts, as well as a resurgence of interest in gardening, organic foods, camping, and do-it-yourself handicrafts. Sales of canned foods have dropped in the face of a renewed craving for fresh produce.

This urge to preserve the best of the past does not mean that the quest for technological improvement will be curtailed. It is worth recognizing that today's backpackers make grateful use of aluminum and nylon, both of which are products of energy-intensive technologies. Engineering is committed to the prospect of new discoveries, and engineers still look eagerly to ever receding horizons. We are tinkerers at heart; we cannot keep our hands off the world. However, the over-optimism, and perhaps even arrogance, that had been creeping into the engineering view is being replaced by a more thoughtful but still enthusiastic commitment to change.

These, then, are what I take to be the main elements of the engineering view: a commitment to science and to the values that science demands—independence and originality, dissent and freedom and tolerance; a comfortable familiarity with the forces that prevail in the physical universe; a belief in hard work, not for its own sake, but in the quest for knowledge and understanding and in the pursuit of excellence; a willingness to forgo perfection, recognizing that we have to get real and useful products "out the door"; a willingness to accept responsibility and risk failure; a resolve to be dependable; a commitment to social order, along with a strong affinity for democracy; a seriousness that we hope will not become glumness; a passion for creativity, a compulsion to tinker, and a zest for change.

The reader may notice that I have not included a staple of much writing about engineering—the urge to serve humanity. Most of the engineers I know are good people but truly no more altruistic than the average citizen, and I feel it is somewhat deceptive for us to imply that this is not the case. Engineers are not missionaries. As professionals we pledge ourselves to public service, but I think this is stating the case somewhat backward. By being hard-working, responsible, dependable, and creative we end up being of service to the community, as well as enhancing our own pride and pleasure. This brings us to the fringes of engineering ethics, a topic I will address shortly. (There is a difference between a general world view and a formal moral code, and for the moment it is the former that concerns me.)

The reader might also notice that I have not referred to the existential pleasures of engineering. This is partly because I have already written a book on that theme and discussed it in a previous chapter. But mainly it is because the existential pleasures—the deep down satisfactions that stem from engaging in the technological work that human beings instinctively want to do—are the reward we receive rather than the goal we seek.

The engineering view is far from the only acceptable way of perceiving the world, and I hope that engineers will be receptive to various types of experience—including the literary, artistic, and political. And, of course, I do not expect all engineers to see our profession exactly as I do. As Santayana has said: "I do not ask anyone to think in my terms if he prefers others. Let him clean better, if he can, the windows of his soul, that the variety and beauty of the prospect may spread more brightly before him."[11]

6

Ethics: (1) Rhetoric and Good Intentions

When engineers get together they do not usually start to philosophize. They are more likely to discuss technical matters—in other words, to talk shop. Next, as with most professionals, the conversation turns to basic microeconomics—how to make a living and protect one's turf. In this regard, things are much the same as they were in the days of the medieval guilds—except that the current agenda includes such matters as errors-and-omissions insurance, pension plans, and government policy toward immigrant engineers.

Nevertheless, just as in the days of the guilds, after the discussion of privilege has run its course, the question of obligation is bound to arise. For what is professionalism if not a mix of privilege and obligation? A profession comes into being on the basis of a special body of knowledge, then builds itself with institutionalized education, accreditation, and self-regulation. Eventually the public grants to the specialists a sort of franchise. (According to some observers the franchise is not so much granted as seized.) In exchange, however, society requires that the experts demonstrate concern for the commonwealth. The professionals, for their part, are sensitive to the expectations of the public, and also respond to their own conscience.

Engineering societies have traditionally treated questions of professional obligation under the rubric of "engineering ethics." The topic receives far less attention than technical and economic concerns, but it makes up in passion what it lacks in priority.

Whenever engineering ethics is on the agenda, emotions come quickly to a boil. During the early seventies, in the wake of Watergate and the environmental crisis, the moral aspect of engineering attracted increasing attention and moved beyond the meeting chambers of practicing engineers into the halls of public discourse.

I consider the subject interesting and important, but the heated feelings it evokes fill me with trepidation. Many are the times that my earnest and well-meaning remarks have aroused the ire of earnest and well-meaning critics. As long as I sing paeans of praise to engineering, I have only friends for readers. As soon as I turn to ethics, however, I invariably find myself in an imbroglio. Yet the subject cannot be avoided, at least not by anyone who seeks to explore the profession in any depth. After we have considered what we do, what we have done, what we think, and how we feel, the question arises, what *ought* we do, how *ought* we behave? And so I return to this thorny thicket where I have struggled many times before.

It is oh so easy to mouth clichés, for example to pledge to protect the public interest, as the various codes of engineering ethics do. But such a pledge is only a beginning and hardly that. The real questions remain: *What* is the public interest, and *how* is it to be served? It is all very well for scholars to opine that an "optimum technology" is one that "is directed towards the highest possible human goals,"[1] or to proclaim that "engineers should make technology more humane."[2] But it is frivolous to suggest that these goals—and the means to attain them—can be readily discerned by good-hearted individuals.

A generalized commitment to do good is worthy; it sets our sights high and puts us in a constructive frame of mind. But too often sermons, codes, and pledges go hand in hand with wrongdoing.

There is a chilling moment in Puccini's *Tosca* when Scarpia, the malevolent chief of police, voices his intention to ravish the heroine and have her lover executed. As he sings of his nefarious schemes, a religious procession appears and the music swells. Scarpia expresses his depravity in an ever more impassioned counterpoint until, just as the procession goes by, he crosses himself and joins in the prayers. This magnificent moment of grand opera reminds us that thus does many a blackguard hide evil intent in sanctimonious ceremony.

Examples of such behavior abound, so much so that expressions of virtue often trigger suspicions of hypocrisy. We have to beware of the cynicism that arises as a response to cynicism. When Samuel Johnson said that "Patriotism is the last refuge of a scoundrel," his friend and biographer, Boswell, hastened to explain that Johnson "did not mean a real and generous love of our country, but that pretended patriotism which so many, in all ages and countries, have made a cloak for self-interest."[3]

Even when there is no question of evil or deception, noble-sounding words can become a soporific, a comfortable substitute for hard thinking, and so work against the cause they claim to serve. Having made this point—which is an obvious one yet often overlooked—the quest for engineering ethics becomes all the more challenging. How is it possible to move beyond platitudes?

In pursuit of the grail of professional morality, I have had a number of illuminating experiences, but none more intriguing than the Second National Conference on Ethics in Engineering held in Chicago in 1982. The conference was hosted by the Illinois Institute of Technology Center for the Study of Ethics in the Professions, and supported by a grant from the National Science Foundation's EVIST (Ethics and Values in Science and Technology). A group of more than a hundred and thirty attended—a few practicing engineers, but mostly teachers of engineering ethics, about evenly divided between philosophers and engineering faculty. Several of the participants carried hot-off-the-press copies of the first issue of the *Business and Professional*

Ethics Journal, with an eye-catching title emblazoned on its cover: "Engineers Who Kill." The new periodical was sponsored jointly by Rensselaer Polytechnic Institute, the University of Delaware, and the University of Florida. Apparently, engineering ethics was very much alive and flourishing, at least in academe.

As William James once said, "Philosophy's queerest arguments tickle agreeably our sense of subtlety and ingenuity,"[4] and during the conference this comment came to mind repeatedly. I found myself enjoying presentations that seemed very queer indeed: for example, the philosophy professor who demonstrated that an engineer's right to refuse immoral orders can be derived from the basic right to behave responsibly, and that this can be shown to rest "on a solid normative base." One of the professor's colleagues commented that this approach was unhelpful, "since it takes us only to the edge of box one."

Another philosopher spoke of the need for engineers to become moral heroes. Claiming that the integrity of each engineer is likely to be put to the test, and alluding to unspecified instances of "spectacular corruption," he insisted that young engineers be prepared to abandon their careers rather than abdicate their moral responsibility. The fact that this feverish picture of engineering life bore little semblance to reality was underscored by the report of two professors who had conducted an investigation of the Dupont Company textile plant in Old Hickory, Tennessee, and the Tennessee Eastman Company in Kingsport, Tennessee. It seems that the engineers interviewed at these two corporations claimed never to have encountered a conflict between their professional obligations and their roles as "company men." Furthermore, they were convinced that such problems as might conceivably arise would stem from lack of competence rather than lack of scruples.

The author of another paper asserted that government regulation could be radically reduced if individual engineers would be "ethically accountable" for the safety of manufactured products.

He did not specify how this could be accomplished in practice.

There was no shortage of such innovative and unworkable ideas. One speaker proposed that the engineering societies be given the main responsibility for regulating the behavior of engineers, disregarding the fact that only a third of American engineers belong to a professional society and that the courts have lately been restricting the power of such societies to regulate the activities of their members. Another speaker suggested that safety engineers within corporations be given "total autonomy."

Finally, as the hour grew late, historian David Noble startled the genteel assemblage with a frontal attack on what he termed "the bogus discipline of professional ethics." He accused engineers of being uneducated elitists and called upon them to unionize and form an alliance with labor to create a "more egalitarian" society. This awoke those of the audience who were beginning to get drowsy and proved a point that should have been obvious from the start, namely, that engineering ethics is less a discipline in which one can expect to see cumulative progress than it is a medium through which individuals can freely express their personal beliefs on almost any topic. (Again William James: "The history of philosophy is to a great extent that of a certain clash of human temperaments.")[5]

By coincidence, two days after the Chicago conference, the Ethics Committee of the American Association of Engineering Societies met in Washington, D.C., trying once again—in an endeavor that is decades old—to come up with a uniform code of ethics that might be endorsed by all the major societies. According to reports of those who were present, the discussions lasted all day and consisted mainly of trying to accommodate the proposed AAES Model Code to a draft prepared by a task force from the Institute of Electrical and Electronics Engineers. A non-discrimination clause proposed by the IEEE group was adopted almost intact. Nothing to argue about there. In another area, where the AAES version said, "Engineers shall act in such a manner as to uphold and enhance the honor, integrity and dignity of the profession," there was added, upon request, "and shall associate only with honorable enterprises." Why not? Even on the ticklish subject of whistle-blowing, tentative agreement was reached by adding innocuous language about engineers, upon discovering a danger to the public, advising "their employer or client and if necessary such public authority as may be appropriate." The "if necessary" discreetly leaves the question open.

But the IEEE draft contained many provisions that were not accepted: an appeal for activism, including supplying "voluntary professional services to worthy causes" and "seeking change through the legal process" (a bit radical for some tastes); an injunction to "disclose" potential conflicts of interest (engineering codes have traditionally prohibited them); a commandment to "neither offer nor accept bribes" (admirable, but a trifle blunt). And much more. Responding to a questionnaire concerning the proposed AAES code, IEEE members were predictably ambivalent. While 72 percent of the respondents agreed with the "fundamental principles," and 65 percent found the code "acceptable overall," only 45 percent were satisfied with the more detailed "canons."[6] In December 1984 the AAES Board of Governors adopted a Model Guide for Professional Conduct, but as could have been predicted, this elicited further critiques from representatives of IEEE and the other societies.[7]

Where will it all end? For the engineering societies there will be more meetings, negotiations, referendums, and perhaps some day an inoffensive unified code, although I wouldn't bet on it. Certainly a code with real meaning and teeth is beyond the realm of possibility. When *Chemical Engineering* presented its readers with a number of "Perplexing Problems in Engineering Ethics," the letters in response were dramatically divided and philosophically widely diverse.[8] Such an experiment is enough to indicate that the prospect of achieving homogeneity seems unrealistic—and, some would say, undesirable.

For academics, there will be more conferences and publications and courses in which everything gets discussed and nothing gets decided. In other words, engineering ethics is going in dialectic circles—which is its saving grace. If I thought it were heading toward formal doctrine I would be concerned, for the prospect of American engineers thinking and acting in concert is a truly Orwellian nightmare.

There is, however, little danger of this. Engineering ethics cannot paper over deep and heartfelt differences of opinion. Inevitably, people of opposing views will each hold high the banner of professional morality, much as opposing armies used to claim that God was on their side.

Nevertheless, the thrashing about, the impassioned and often silly debates, the search for virtues that nobody can define—these sharpen our wit and enrich our spirit. Therein lies the value of anarchic ethics conferences, confusing ethics classes, and amorphous ethical codes. In engineering it is often the process rather than the product that captures our interest. Nowhere is this more true than in the field of engineering ethics.

It is too bad that the "Boom in Ethics," as *U. S. World & News Report* called it in 1976, has begun to dissipate. In the budget-cutting frenzy of 1985, the National Science Foundation came close to eliminating the EVIST program that sponsored the Second National Conference on Engineering Ethics. A Third National Conference, entitled "Engineering Safety: The Value Dimensions of Controlled Hazardous Technology," was held in May of 1985, but funding for more such gatherings may be hard to come by. Although such conferences will not solve our problems in any definitive sense, a refusal to support them implies that we do not care and that the interest in professional morality was only a passing fad.

There are other disturbing symptoms. Surveys of members of the American Society of Civil Engineers showed that from 1976 to 1984 there was a marked drop in concern for a code of ethics.[9]

In 1985 the IEEE's Society on Social Implications of Technology suffered a precipitous drop in membership. This trend is no less alarming for being in tune with the times. Surely we engineers do not want the public to get the impression that we are losing interest in the moral aspects of our profession. More important, we must not permit such an idea to take hold in our collective psyche.

7

Ethics: (2) Illusion and Change

Although engineers have irreconcilable differences of opinion on many important matters, there may still be some meaningful ethical principles on which they can agree. A time-honored way to approach a problem in engineering—and in philosophy as well—is by process of elimination. What if we were to apply this method to engineering ethics? If we peeled away each of the concepts that ethics is *not,* we would eventually be left with the real thing.

I have already noted that morality cannot be achieved by making noble-sounding proclamations—notwithstanding the value of nurturing noble sentiments. I have also rejected some of the more fantastical proposals of academic philosophers. The morality we seek should be founded in the real world and be capable of practical application. It should be able to meet standards of both good will and good sense.

To continue: I suggest that engineering ethics is not a set of guild rules. Once upon a time that is all that engineering ethics was. The original codes of ethics told engineers how they should deal with each other. Don't compete for commissions on the basis of price. Don't advertise. Don't review a fellow engineer's work without first getting clearance from him. For many years so-called violations of ethics were invariably breaches of such rules. Today we might call such admonitions professional protocol, professional manners, or professional decorum. But society at large will not accept them as having anything to do with real ethics. In fact, society at large, acting through the Department of Justice and the courts, has declared such restrictions to be in violation of antitrust laws. This occurred during the 1970s and radically changed the status of all professions—not only engineering. What engineers used to call ethical behavior is now to a great extent illegal.

I hope and trust that engineers will continue to be gentlemanly—or ladylike as the case may be. Certainly, cutthroat competition in engineering is not in anybody's best interests, no matter what the antitrust laws say. But it is a fact that engineers can no longer hope to control their own profession totally from within. The law will not permit it, and in a democracy this is probably fitting as well as inevitable.

The law has taken over many substantive areas that used to be the province of professional ethics. Confidentiality, industrial secrets, and proprietary interest in invention are defined by statute and legal precedent rather than by ethical codes. So are the norms of professional responsibility. For example, an engineer in New York paid invoices from a firm that had performed no services for him, and when it was discovered that this was done to make a covert political contribution, the Commissioner of Education suspended the engineer's state license, an action that was upheld by the Appellate Division of the New York Supreme Court. In Florida, a structural engineer's license was revoked after the collapse of a building he designed. He was found guilty of "negligence, incompetence, and misconduct in the practice of engineering . . . in violation of Section 471.033(1)(g) Florida Statutes."[1] Among other things, this engineer had affixed his seal to incorrect design calculations without bothering to supervise or check the work of his subordinates. He also approved installations in the field without inspecting them as required by code.

Naturally, it is unethical to break the law. But is this worthy of extended discussion? Need we spend time teaching young engineers not to break the law? Bertrand Russell once observed that some of the most interesting questions of classical philosophy have now been absorbed into modern science and hence no longer require the attention of philosophers. I believe the same thing has happened with engineering ethics and the law. Once an ethical imperative has been translated into a legal injunction we should be able to consider the matter settled. Will an engineer in the course of a career ever be asked to break the law? Rarely, but it might happen—just as it might happen to an accountant, a warehouse foreman, a health department food inspector, or a president of the United States. The answer is the same in each case: Do not break the law! There is no harm in repeating this occasionally, but such advice is not worthy of a prominent place in the literature of engineering ethics.

Concepts of product safety also are moving out of the realm of ethics. Increasingly they are being absorbed into the province of government regulation. There was a time when the rule of the land was *caveat emptor,* buyer beware, when Congress believed it was prohibited by the Constitution from telling manufacturers how to manage their affairs. In the 1830s, Senator Thomas Hart Benton, arguing against legislation intended to improve the safety of steamboat engines, proclaimed that the proper way to tell if a boat was safe was to personally inquire as to whether the machinery was in good working order.[2] The "good old days" were really wild, woolly, and relatively lawless days. Robber barons singlemindedly pursued profit, and a swarm of rapscallions, exemplified by the snake-oil salesman, plied their wares with little or no restraint. In such a climate, a conscientious engineer was sometimes the only protection the public could look to. Discussions of engineering ethics often revolved around the question of how engineers could maintain their independence from industry in order to safeguard society.

The ethics of those relatively primitive times, however, will not serve as a model for today. As the nation became wealthier and better educated, the communal will decreed that laws be passed and rules and regulations established to defend the public interest. In 1852 Congress enacted a law that required the inspection of steam boilers. In the 1880s it passed railroad safety legislation. Between 1880 and 1906, 103 bills were introduced proposing to control interstate traffic in food and drugs, until finally the Pure Food and Drugs Act was passed and signed by Theodore Roosevelt. The Federal Trade Commission Act was passed in 1914, the Federal Power Commission founded in 1920, the Federal Communications Commission established in 1934, and the Civil Aeronautics Board in 1938. Starting in 1970 there was a dramatic increase in government concern for health, safety, and environmental protection. A listing of some of the better-known protective measures tells the tale: Occupational Safety and Health (1970); Boat Safety (1971); Insecticide, Fungicide and Rodenticide (1972); Marine Protection, Research and Sanctuaries (1972); Consumer Products Safety (1972); Noise Control (1972); Safe Drinking Water (1974); Motor Vehicle Safety (1974); Hazardous Materials Transportation (1975); Toxic Substances Control (1976); Solid Waste Disposal (1976); Clean Air (1977); Mine Safety and Health (1977); Water Pollution Control (1977); Comprehensive Environmental Response, Compensation and Liability (Superfund) (1980).

The proliferation of these laws, and the agencies to administer them, is a manifestation of the public will and a splendid example of democracy at work. The legislation was not intended as an insult to engineers. In fact, engineers participated in drafting many of the acts that today protect the body social from disease, injury, and disaster. I do not understand why so many people—engineers and philosophers among them—consider the growth of government regulation as a defeat for professional morality. I particularly do not understand why so many business executives think this phenomenon inimical to economic and technological progress. It is really a shame—and a great danger—that government regulation has, in our time, gotten such a bad name. We cannot expect to enjoy the benefits of high technology and the fruits of a competitive economy without facing up to the complexity—yes, the annoying complexity—that is inherent in them.

The industrial culture of which we are so rightly proud, and in which most engineers spend their professional lives, is called the world of free enterprise—a euphemistic way of saying that it is something of a jungle. This fierce place is an outgrowth of the Darwinian jungle—nature red in tooth and claw—and also of thousands of years of ferocious human competition—wars, revolutions, invasions, and crusades. But the jungle is not just a remnant of our savage past. It is part of the carefully nurtured present, and our society goes to great lengths to protect it. We say that competition is good for the consumer. We say that competition nurtures innovation and promotes productivity. In America we are determined to foster economic competition at almost all costs. If executives from different corporations are caught coordinating their price lists we send them to jail.

We recognize that this is not the ideal world of our dreams. The glorification of profit-seeking makes us uneasy. The tragedy of bankruptcy and unemployment is appalling. We deplore the extremes that we call "price wars." We allow a certain amount of controlled monopoly, and we permit some cooperation between competitors. We even subsidize certain industries. But we do this uncertainly and grudgingly. The jungle brings such wonderful benefits to our society that we are determined to preserve it.

Unfortunately, total freedom and pure competition—intense life and death competition—evoke a reckless expediency and occasional cheating, and in turn this can lead to shoddy merchandise, dangerous products, depleted resources, and a defiled environment. We attempt to deal with this problem by establishing rules and regulations. Naturally, as our technology becomes more advanced and our society becomes more complex, our rules and regulations become more numerous and our means of enforcing them more various. The proliferation of rules and regulations is one of our ingenious adaptations to jungle living, and is one of the glories of our civilization.

Of course, the rules are imperfect and the administrators of these rules are often foolish and overbearing. We can agree that regulations must be made more sensible and agencies more effective, but this is not the same as saying that if engineers were more ethical the government would not have to create so many laws and agencies.

The concept of industrial self-regulation does not bear scrutiny. When self-regulation is not enforced by some form of law it leads to the prosperity of those who choose not to join the responsible group. It rewards the renegades and rascals. The self-discipline of a cautious and responsible citizen is not binding on his unscrupulous neighbor. Consumers are always ready to complain, but they have shown little inclination to reward well-intentioned businesses by paying more to buy their products. And nobody has suggested that corporations or engineers should be granted special permission to police their own industry. We do not countenance the punishment of people unless they have violated a law. To the contrary, we protect the rights of individuals against the pressure of their peers. In business we do this in the name of free enterprise. So self-regulation is reduced to the granting or withholding of the equivalent of a Good Housekeeping seal of approval.

Even more telling, if society were to rely upon self-regulation, according to what criteria would it be carried forth? We all want safety, reliability, an undamaged environment, and undepleted resources. We also want economy—it is a moral imperative in that it means more benefits for people of modest means. By what mechanism would corporations and engineers decide on the many trade-offs that must be made, and why should the basic decisions that affect our lives be left to an unelected unrepresentative group? Do we really want the automobile companies, for example, or the engineers in their employ, to decide in their wisdom whether a car should be safe when hitting a solid wall at 30 miles per hour, 35 miles per hour, or 135 miles per hour? The public pays for safety (not the corporations, who will make their profit in any event). So the public—through its representatives—should decide how much safety it is willing to pay for.

It has been suggested that engineers may be pressured to violate ethical principles by managers who favor profit over public safety. In most cases, however, the problem is not a matter of ethics, but a matter of reasoned choice—not what risk is moral, but rather what risk is *acceptable,* and this is a choice, I maintain, that should be made by society as a whole. If there are laws and regulations—if there are standards and guidelines—then engineers and managers should have no reason to be at odds with each other. Where they do differ, they will differ as individuals, not as ethical engineer versus amoral manager.

Our society has painstakingly developed more than twenty thousand voluntary technical standards—national standards—and these constitute an engineering triumph—another glory of our civilization, comparable to the development of our regulatory laws and agencies. In fact, these two marvelous phenomena are interrelated. Standards are developed by professional groups and then government agencies, when they see fit, adopt the standards and give them the force of law. Without these laws, regulations, codes, and rules, each engineer would be given unwarranted—and unwanted—powers (and, incidentally, each engineering problem would entail reinventing the wheel).

We must constantly work to make our regulatory apparatus more efficient. This is a special challenge for engineers who choose to go into government service. And we must constantly preach self-discipline. Even with laws we need self-discipline in the form of scrupulous compliance. But self-discipline is no substitute for restraint by government. The American founding fathers knew this and said it—frequently and eloquently.

Of course, it would be undesirable and impossible to subject every engineering decision to a written regulation. The next line of defense, however, does not then become the ethics of engineers, but rather standards of liability as made manifest by our courts. How safe is safe enough? What is the *expectation* of the citizenry in the area of risk and responsibility? When specific regulations do not apply, it is the decisions of judges and juries that best indicate what is appropriate. We do not say to industry, "be ethical." That is too vague and permits an enormously wide range of individual response. It is an invitation to chaos. We say instead, "be prudent, and here are the standards by which you will be judged."

An industrial firm can be held liable for defective product design, defective manufacture, inadequate labeling, and faulty packaging. Liability losses can be suffered by companies that fail to keep proper records of product sales and distribution, or fail to keep adequate records of project failures and customer complaints. As stated by the authors of *Product Safety and Liability: A Desk Reference,* "The one choice engineers, designers, and their employers no longer have is whether or not to pursue safety goals. To ignore or pay only lip service to safety . . . can jeopardize corporate survival."[3]

Product safety is no longer a matter of conscience and good will. It is a professional specialty founded in knowledge of prediction techniques, fault tree analysis, failure modes, effects analysis, and the operator-design interface. Some experts have estimated that the cost of poor quality can amount to 25 percent of sales at manufacturing companies and up to 40 percent of operating costs at service companies.[4] The American Society of Safety Engineers reports that its membership is increasing. Corporate ombudsmen solicit employee warnings.

When it comes to establishing liability, worthy intentions no longer count. Tort law traditionally required that negligence be proved as a "proximate" cause of an injury, but since the 1960s courts have been imposing "strict liability" for injuries caused by dangerous products *whether or not* negligence was involved. The situation has progressed to the point where corporate executives now claim to be the aggrieved parties. (For example, an overweight man who suffered a heart attack trying to start a lawnmower won a $1.75 million award against Sears, Roebuck & Company on the ground that the starter rope was too hard to pull.) Industry leaders have appealed to Congress for protective laws, maintaining that the national economy is adversely affected by the ever stricter standards of accountability. (For example, the FMC Corporation declined an opportunity to produce precision bearings for helicopter rotors because the potential profit did not justify the liability risk if the helicopters crashed.) Insurance companies play an increasingly crucial role in business planning.

For consulting engineers the situation has reached crisis proportions. In 1985, insurance industry statisticians reported that such engineers have a 43 percent chance of having a claim made against them and that over half of these claims will result in a serious financial impact on the practitioner.[5] Insurance rates have risen precipitously. In 1986 consulting engineers were paying 4.1 percent of their gross billings for liability coverage, an increase of almost 50 percent in one year.[6]

Awards against industrial polluters are also mounting. Where government enforcement efforts are deemed insufficient, public-interest groups have stepped in. Most of the major environmental protection laws contain provisions permitting private suits, and hundreds of these have been instituted, particularly under the Clean Water Act.

However Congress and the courts eventually decide to apportion the costs of injury and environmental degradation, it is apparent that engineering ethics is not the main issue. At Quality College, a California institution whose program attracts hordes of anxious executives, four "absolute" commandments are posted on the wall. The first one states: "Quality has to be defined as conformance to requirements, not as goodness."[7]

* * *

Finally, I suggest that engineering ethics is not, or should not be, a medium for expressing one's personal opinions about life. I touched on this briefly in the previous chapter, but the matter goes far beyond the emotional pronouncements of speakers at seminars.

There are serious issues before us: safe products versus economical products; protecting the environment versus exploiting our resources; armament versus disarmament. Shall we explore space or use our money for supporting libraries or building day-care centers? Shall we use pesticides that save lives at the risk of doing damage in the food chain? Shall we drill for oil offshore at the risk of polluting our beaches? Shall we build a dam or protect an endangered species? These are essentially political issues, matters of choice for society, and for the most part they should be decided by the actions of officials we elect, by laws, by regulations, and by decisions in the courts, even occasionally by referendum. They certainly can't be resolved by engineers promising to be moral.

Engineers do not have the responsibility, much less the right, to establish goals for society. Although they have an obligation to lead, like most professionals, they have an even greater obligation to serve. I may wish to build a cathedral or a hospital, but if I am commissioned to build an office building, or even a gambling casino, it is professionally proper for me to accept the commission. Naturally, I have my own conscience to contend with. A few years ago my company declined an invitation to bid on the construction of a building for the government of Libya. This was before that government was known as a sponsor of international terrorism, but it was reported to be sending assassins all over the world to track down dissident expatriates. These were bad people and we wanted nothing to do with them. However, if all American engineers refused to work for the Libyan government—if engineers as a group engaged in a political boycott—would that be a good thing? I do not think so, particularly if the United States government approved of the building in question, believing that efforts should be made to encourage trade and bring that renegade government into the family of law-abiding nations.

Let me use another example from personal experience. My wife and I have a weekend cabin by a small lake north of New York City. A few years ago the State Parks Commission decided that they would develop a lakeside beach, with bathhouse and parking lot, in a nearby public park. The homeowners association to which we belong retained an engineer to study the possible adverse effects that this development might have on our lake. A dispute ensued that was finally resolved when we got some concessions concerning sewage treatment and traffic control. For all I know, our engineer thought we were a bunch of elitist snobs trying to make it difficult for other people to share the pleasures we enjoy. And perhaps the state's engineer was a nature-lover who thought it a shame to develop a public beach in our mini-wilderness. But assuming this was the way they felt, I do not believe that these engineers acted immorally in working for causes they did not espouse. They were giving professional advice to people who were entitled to access to professional advice.

It is marvelous, of course, if you can work for a cause in which you believe deeply. I hope that many young engineers will devote themselves to public service or otherwise become involved in work to which they can give their whole heart. But the world being the way it is, this is not always possible, and the average engineer has no reason to feel apologetic or defensive about it.

Thinking along these lines, we come inevitably to the subject of armaments. Is it not unethical to design or be in any way involved with weaponry? I like to think of engineering as a life-enriching profession. Yet I recognize that technology can be applied to any human activity, and war—or let us say national defense—has been a matter of vital concern to practically every society throughout history. Unless I am a pacifist and believe in unilateral disarmament—which I am not and do not—it would be hypocritical of me to accuse engineers in the defense industry of being unethical. Are soldiers unethical? Are the members of Congress unethical when they appropriate funds for the military? Are citizens unethical when they vote for these representatives and support their policies? Much of the human race, I fear, shares responsibility for the dreadful dilemma of armaments.

Two-thirds of the engineers who work for the federal government are employed by the Department of Defense. Thirty percent of *all* American engineers perform work that is financed by the federal government, and a very large portion of this work relates to defense.[8] In reply to a questionnaire prepared by the Institute of Electrical and Electronics Engineers, 68 percent of the respondents said that they had at some time worked on defense technology. Thirty-nine percent said they would prefer not to engage in such work, and if the weapons were nuclear the figure increased to 48 percent. On the other hand, 56 percent indicated they would be proud to work on weaponry; for nuclear weapons this figure dropped to 42 percent.[9]

I do not see how one can conclude that all engineers have an ethical *obligation* not to work in the defense industry. On the other hand, engineers do have an ethical *right* to make such a choice as a matter of personal conscience. In 1981 a group called High Technology Professionals for Peace was founded in Cambridge, Massachusetts, its avowed purpose being to find non-defense-related jobs for its members while also educating the public about the threat of nuclear war. At about the same time, several electrical engineers in Stanford, California, formed Beyond War, an association devoted to finding ways of eliminating war and weaponry. Such organizations provide a means for like-minded engineers to express personal feelings about the uses to which engineering is put, feelings that usually cannot—and ought not—take precedence in the workplace.

It is perfectly proper—let me say it again for emphasis—perfectly proper for engineers as individuals to take political stands on technical matters. In the summer of 1985 several hundred scientists and engineers at Cornell and the University of Illinois, and subsequently on other campuses, signed a pledge not to work on "Star Wars" research and urged colleagues to join them in this stand. At the same time the office of the Strategic Defense Initiative of the Department of Defense reported that it had received twenty-six hundred applications from individual researchers and universities anxious to participate in the program. When the IEEE sponsored an Electronics and Aerospace Systems Conference, several of its members protested that papers critical of the Strategic Defense Initiative were not accepted for presentation on the program, and the matter was debated in the pages of the organization's journals.[10] Debate is the very essence of a democratic society, and this applies to engineers as well as to anybody else. However, the line between wholesome argument and ruinous divisiveness is often exceedingly thin. As members of the U. S. Senate are wont to say when the debate becomes excessively acrimonious: "We are Americans on both sides of the aisle." Just as a great nation must cope creatively with controversy, so must a great profession guard against corrosive internal strife.

Inevitably someone will raise the example of Nazi Germany. Are not engineers at some point personally and organizationally responsible for their actions? To which the only reply is that, of course, each individual is in the end accountable to an inner moral code. But it is neither fair nor rational to compare the death camps of a malevolent dictatorship to the defense industry of a democratically constituted republic.

Armaments are not the only morally ambiguous work in which engineers are engaged. I would not like to be called upon to design a cigarette-manufacturing machine. Yet I cannot condemn those engineers who do so unless I am also willing to condemn tobacco farmers and the many thousands of my fellow citizens who sell cigarettes in grocery and stationery stores. It is the ultimate cop-out to blame individuals, including engineers, for actions condoned by the commonwealth. And it is the ultimate folly to rely upon individual preference in matters that require communal decision.

In some respects the obligations of an engineer remind me of the obligations of a judge. Irving R. Kaufman of the United States Court of Appeals for the Second Circuit has stated the case as clearly as I have seen:

> In his decisions, a judge must be obedient both to the commands of law and to the dictates of the Constitution. We are not the arbiters of what, in our view, should be. We are the interpreters of what is. Our role involves creativity and discretion, but it is imagination of an interstitial character that is required. When faced with a contrary legal mandate, a judge has no choice but to put aside his personal policy preferences.[11]

Robert H. Bork, another U. S. Appeals Court judge, has said that "the morality of the jurist" consists of the "abstinence from giving his own desires free play . . ."[12] Justice John Paul Stevens of the United States Supreme Court puts it more bluntly: "It is not our job to apply laws that have not yet been written."[13]

I suggest that in engineering there should prevail a doctrine equivalent to what in law is called "judicial restraint." It is not the engineer's job, *in his or her daily work,* to second-guess prevailing standards of safety or pollution control, nor to challenge democratically established public policy. As a citizen, to be sure, he may work for modification of such standards and policies. Judge Kaufman makes this clear about jurists as well:

> I do not wish to be misunderstood. I am not cautioning against recommending candidates who have taken sides in national debates on pressing issues.

> Such participation does not augur bias, but rather a dedication to the commonwealth that should be encouraged in all public officials, judges included.[14]

Jurors, too, are traditionally admonished not to be swayed by personal sentiment but rather to decide upon the facts and then to apply the law as it is interpreted by the court. That is how society expects jurors to do their duty.

Does all of this mean that the engineer is to serve his employer or his client "totally," bound only by the limits of established rules? Does it settle the matter to assume that the public interest is protected by the rules and the courts?

Let us seek another analogy in the third leg of the judicial triangle, the practicing attorney. Judge Kaufman, attending a symposium sponsored by the President's Commission on Organized Crime, was dismayed to learn of the critical role played by lawyers in organized-crime syndicates. He speculated that this stemmed from the doctrine, taught in law school, that the lawyer's role is, above all, to serve his client. "But," Kaufman concluded, "unrestrained and overzealous loyalty to a client never is appropriate. Whoever the client, a lawyer always must serve another interest as well." And what is this other interest? "The law . . . the welfare of the legal process."[15] Thus, in admonishing attorneys not to go beyond proper limits in serving clients, he does not direct them toward something as amorphous as "the public interest," but rather toward a vital social institution that has been placed in the care of the legal profession—the law as a philosophical construct; the legal process.

What comparable social institution has been placed in the care of the engineering profession? After we have peeled away and discarded rhetoric, unworkable ideals, guild rules, illegality, self-regulation as opposed to guidelines established by the government, individual opinion as opposed to standards of liability evolved in the courts, whim and personal world view, what is the precious essence that remains? Let us look further.

8

Ethics: (3) Credo for a New Age

It is difficult to acknowledge that the world has changed and that old verities may be subject to challenge. The concept of goodness survives, of course, but the way in which it manifests itself has been transmuted. Not long ago the cowboy in a white hat, walking down Main Street with guns at the ready, was a hero. Today he is a vigilante. Frontier conditions no longer prevail.

A similar change has occurred in the world of engineering, and that we have been slow to perceive it is surely understandable. Most of the changes I alluded to in the last chapter—antitrust restrictions on professional independence, vastly increased government regulation, broadened concepts of legal liability—have taken place so recently that their significance has not yet become clear. The startling fact is, political and legal reforms have made much of the old engineering ethics moot. When Edwin T. Layton, Jr., a respected historian of technology, calls for engineers to become a "loyal opposition" within corporate America, he demonstrates his failure to understand the nature of the new social order.[1]

I propose that the essence of engineering ethics be recognized as something different and apart from white-hat heroics. There is no single word adequate to define this "something," but *conscientiousness* is fairly close to what I have in mind. In Chapter 5 I noted that engineers are accustomed to having their fellow citizens rely upon them. As a consequence, if engineers do their jobs well they are being more than competent; in serving their fellow humans, and serving them well, they are being good. Reliability is a virtue. Therefore, the conscientious, effective engineer is a virtuous engineer.

If we relate engineering ethics to protection of the public interest, then clearly diligence is more moral than conventional "morality." A kindly, generous, well-intentioned, even saintly engineer may still be an inept engineer, that is, a "bad" engineer whose work does not serve the public well. Engineering ethics is not the same as conventional ethics. In technical work, competence is more good than "goodness." In support of this somewhat paradoxical view let us look at some facts.

In 1982 there was founded at the University of Maryland the Architecture and Engineering Performance Information Center (AEPIC). Its initial funding came from a National Science Foundation grant, and subsequently it was supported by the university and corporate sponsors. The purpose of the center is to collect, classify, and analyze examples of technological failure. Data has been amassed on some forty thousand instances of structural, electrical, mechanical, aesthetic, or environmental failure in buildings and other civil works. A noted forensic engineer who has reviewed a sampling of this data and compared it with his own lifetime of experience reports that "probably less than two percent" of the failures relate to unethical behavior in the conventional sense, that is, greed or intent to deceive, and even these cases usually contain a quotient of sloppiness.[2] Engineering avarice usually manifests itself in a reluctance to spend time (which is money), an avoidance of doing things over again.

Researchers at the Swiss Federal Institute of Technology in Zurich analyzed 800 cases of structural failure in which 504 people were killed, 592 people injured, and many millions of dollars in damage incurred. They concluded that "the causes of failures should be seen in human mistakes, errors, carelessness and so on. . . ." Where engineers were at fault, the "types of human unreliability" were itemized as follows:

	percent
Insufficient knowledge	36
Underestimation of influence	16
Ignorance, carelessness, negligence	14
Forgetfulness, error	13
Relying upon others without sufficient control	9
Objectively unknown situation	7
Unprecise definition of responsibilities	1
Choice of bad quality	1
Other	3
	100

The findings indicated that most of the mistakes could have been detected in time by additional supervisory control. Even without additional control, however, "roughly one half of the failures could be avoided just by critical inspection of data, plans, etc. . . . by the person next involved in the planning and construction process."[3]

These statistics only bear out what we know through common sense and daily experience. Practically all disasters—think of Bhopal and Chernobyl, the two most dreadful of recent years—are attributable to one or another type of ineptitude.

One can only conclude that the great need in engineering ethics is an increased stress on competence. This is the human endowment, the vital social institution, that has been placed in the care of engineers, just as legal process has been entrusted to lawyers. In fact, the concepts of technical competence and political justice are related. As John W. Gardner wrote in his 1961 book, *Excellence:*

> The importance of competence as a condition of freedom has been widely ignored (as some newly independent nations are finding to their sorrow). . . . Keeping a free society free—and vital and strong—

> is no job for the half-educated and the slovenly. Free men must be competent men.[4]

Other words come to mind: dedication, energy, self-discipline, caution, alertness, awareness—and most of all, as I have suggested, conscientiousness. The greatest threats to moral engineering are carelessness, sloppiness, laziness, and lack of concentration. An engineer may start out honest and high-minded but become immoral by falling prey to one or more of these sins. On the other hand, an engineer who starts out by being conscientious must end up by being honest, since competent engineering, excellent engineering, is in its very nature the pursuit of truth. A conscientious engineer, by definition, cannot falsify test reports or intentionally overlook questionable data, cannot in any way evade the facts.

I do not suggest that engineers should be "blindly" conscientious. No whole person will do anything blindly. A principal purpose of this book is to encourage engineers to be introspective. But I do wish to call attention to evidence that I consider intriguing and important. This evidence, as reviewed briefly above, shows that society's great need is for competent engineers rather than self-righteous ones. Even in the morally ambiguous field of weaponry, there is a crying need for competence. Perhaps the greatest danger to humanity is that devastation may come by way of technical error—not by evil intent, but by *miscalculation.*

While the headlines speak of hawks and doves, our best hope for survival may lie with the "owls," those analysts who argue for improved technical systems that can reduce the chances of a major nuclear war. For example, the United States and the Soviet Union could place unmanned tamper-proof sensors in each other's missile fields in order to increase confidence in the validity of early warning signals and reduce the chance of either nation reacting to a false alarm. It would also be helpful for potentially hostile nations to jointly establish rapid, reliable, redundant, and survivable systems of communication.

* * *

It is not enough for us merely to avoid technological disasters, important as that may be. An equally pressing objective is to use the world's resources so as to best meet the needs and aspirations of the global population. Here political and economic issues are at least as vexing as technical problems. But what society requires from the engineer, again, is conscientiousness.

Consider a civil engineer I know who is remarkably adept at designing economical flat-plate concrete slabs for apartment houses. With careful study and hard work (always staying within safety factors prescribed by code), he is able to develop plans in which the reinforcing steel used is as little as four pounds per square foot, whereas an average plan might run over five pounds per square foot. I also know an engineer who designed a comparable building that turned out to have seven pounds per square foot. It was a hurried, lazy, unimaginative piece of work, and when I pointed out to him how costly his design would be he merely shrugged. He could not be bothered to invest the additional effort in a more cost-effective design. The difference between the two designs represented 3 percent of the total cost of the building, and since the projects I refer to involved housing for the indigent elderly, the harm done by the lazy engineer, or the good done by the diligent engineer, is self-evident. This is an aspect of engineering ethics that is too often overlooked.

Concrete slabs in housing projects may seem a mundane consideration. But imagine effective engineering applied to all the world's material needs: energy, food, clothing, and the rest. Seen in this context, an improvement of 3 percent becomes monumental, and an engineer's failure to do his or her best becomes a sin against humanity.

At this point something beyond care and diligence enters the equation—let us call it *inventiveness.* Our concept of conscientiousness must be broadened to include innovative thinking. In a world of hunger and destitution, the creative engineer, along with the dedicated engineer, becomes a towering force for moral good. The oft-repeated statement that genius is nine-tenths perspiration testifies to the importance of pure effort in the inventive process. As for the final tenth—the elusive quality of inspiration—it, too, is associated in our minds with goodness.

It is interesting to consider the ways in which competence and virtue are intertwined in our thinking. We say "good job" when we actually mean "well done job" or "competent job." We use "conscientious" to describe a person who is diligent and hardworking, yet the word literally means "governed by the dictates of conscience." There is a venerable philosophical tradition that equates goodness with truth, a tradition that stems from the ancient Greeks, who thought the search for facts was a form of virtue.

Practically speaking, people who start out to do a good job and people who start out to do good often end up in the same vicinity. Many different journeys can lead to the same shrine. I have debated ethical theory with certain individuals only to find that, given a real-life situation, we advocated the same resolution. Often, ethics is merely a question of emphasis rather than substance, not that this makes it unimportant. For poets, clerics, ethicists, politicians, and educators, there are a variety of moral roads to be followed and much to be said in favor of each. For engineers, however, the one true road must pass through effectiveness and creativity, and the ideal starting place is conscientiousness.

There still remains the question of whether engineers, as the agents of technology and of technological change, do not have some unique responsibility for the direction that technology takes. If there are too many factories in the landscape, too many TV screens in our homes, too many chemicals in our food, too many weapons in our arsenals, is there not always an instinct to blame the engineers? I have argued that these are essentially political problems for society as a whole. Nevertheless, by stressing the importance of political decision-making in a republic, I do not mean to underestimate the special contributions—beyond doing a good job—that can, and should, be made by professionals.

A most productive way for engineers to apply their talents to communal needs is in volunteer activities generally called *pro bono.* Although engineers are not uniquely qualified to plot the technological course for society, they *are* uniquely qualified to explain how technologies work and what their consequences may be. So part of the engineer's responsibility to society should be discharged in education.

Engineering societies have long recognized an obligation to commission studies, issue reports, set standards, schedule seminars, warn, lobby, and cajole. They have a quite respectable history of calling attention to technological problems: overexploited resources, polluted air and water, deteriorating infrastructure, and so forth.[5] They engage in such activities not only from altruistic motives but also because every new technological difficulty is a new engineering challenge, that is, a potential engineering job. Cleaning up oil spills, storing radioactive waste, purifying groundwater, preserving wetlands, designing disarmament techniques—coping with the adverse effects of technical progress—is just as much engineering work as drilling wells, building power plants, manufacturing chemicals, or developing weapons systems. Thus it is difficult to separate activities that are public-spirited from activities that further the cause —and the employment opportunities—of the profession. Overall, however, I believe that the societies work hard at being good citizens. William Lowrance has called upon professionals to provide "stewardship beyond narrow responsibility,"[6] and, however imperfectly, the engineering societies endeavor to respond to this summons. Unfortunately, no more than a third of the nation's engineers belong to a professional organization.[7]

For the individual engineer, the opportunities are more limited and the responsibilities are less clear. Ideally, all engineers should first join one or more professional societies and play an active role in them. Then, engineers should serve on community boards, write letters to the editor, testify at public hearings, and generally contribute their talents to worthy causes. They should communicate with engineers in other lands and work toward international trust and understanding. They should in every way bring honor to their profession and make technology a force for good in the world. But here I am allowing myself to be transported into the realm of wishfulness. We may in theory make voluntarism a moral imperative. Yet, as a politician has said, we campaign in poetry but must govern in prose. Fine sentiments and exhortations do not often prevail over the pressures and temptations of daily life.

It is the character and competence of individual engineers that will in the end make all the difference. Once we recognize this, we will devote less energy to composing codes of ethics and more to figuring out how best to educate the engineers of the future. We must create for ourselves an ideal—the civilized engineer of the coming era—and then consider ways of making reality out of concept.

At this point, however, I do not want to leave the subject of engineering ethics without making special reference to those engineers who work in the non-profit sector—in academe, public interest organizations, and government. These engineers do a job and earn a salary, and in that sense cannot be deemed more ethical than their counterparts in industry or private consulting. Yet they do work more directly—more intimately—on behalf of the public than do their fellows, and often at considerable financial sacrifice. Without granting them superior moral status, I would like to give them a modicum of special recognition, a tip of the hat, as it were.

Teachers of engineering transmit the profession's heritage from one generation to the next, frequently with enthusiasm, skill, and flair. Engineers in public interest organizations play an increasingly important role and are able to bring much-needed facts and logic to debates that are needlessly acrimonious. (Ralph Nader should have been an engineer!) Engineers who work for the federal government make incalculable contributions in the fields of defense, space, scientific research, national economy, and public health.

Finally, the engineers who have made their careers in public works are the unsung heroes of a little-known but glorious saga. How the teeming, putrifying, corrupt cities of nineteenth-century America were turned into the functioning metropolises of today is a tale waiting for its master historian. We take it so for granted—the water, sewers, paving, transport, garbage disposal, zoning, building codes, parks, and the rest. But it took many brilliant, tenacious, visionary engineers to make it all work, and continue to work, every incredible crisis-filled day. A number of public works engineers have been moral giants, fighting vested interests and garnering support from enlightened citizens' groups. Most have been solid, reputable proponents of the communal good. All, however, have needed to leaven their professional ethics with an abundance of common sense and political acumen, for even in a democracy, rectitude and loftiness of purpose are not in themselves adequate tools for dealing with the public.[8]

Conscientiousness entails dedication—not only or even mainly to an abstract public good—but dedication to the cause at hand. This raises interesting questions concerning the meaning of loyalty, and it is to these questions that I would like to turn next.

9

Loyalty, Or Why Engineering is Sometimes Like Baseball

"What should I do if I am assigned work that I think is not in the best interest of society?" The young man who asked me this question was an engineering student soon to graduate and embark on a career. He was frowning and looked deeply troubled. Even in this time of conservative self-interest, I have met a number of young engineers with similar misgivings.

When I was in school we didn't worry about such matters. We had concerns aplenty, but they were of a different kind: What sort of work did we want to do? What were we good at? How could we best achieve success, whatever that was? As for *loyalty,* it never occurred to us that conflicts could arise. We assumed that an engineer was loyal to superiors, colleagues, and clients, and that service to the community came naturally as the result of a job well done.

Today, in the wake of the environmental crisis and widespread doubts about technology, such a view smacks of insensitivity. If engineering projects can turn out to be harmful—to the environment, say, or to public safety—then it appears possible that one's professional work might run counter to one's obligation to society. Which gives rise to the question put to me by the frowning student.

I did not think it appropriate to launch into the arguments to which I have devoted the past several chapters. This young man was thinking intently about his own personal future and his own very personal feelings. After a moment's thought, I assured him that he was unlikely to encounter the moral crisis he feared. I could say this with some confidence because every engineer I have ever met has been satisfied that his work contributes to the communal well-being, though admittedly I had never given much thought to why this should be so. As I pondered the question, reflecting on the way most people manage to blend personal morality with loyalty and commitment, I found myself thinking—of all things—about the trade of Eddie Stanky.

In the late forties Eddie Stanky played second base for the then Brooklyn Dodgers. He lacked outstanding talent but was known as a great competitor. He was feisty and inventive. He was particularly adroit in reaching first base by managing to be hit with pitched balls. One day I was horrified to learn that this dirty player, this bad sport, had been traded to the team I rooted for, the New York Giants. But then a strange thing happened. Once he was playing for my team he seemed morally reformed. We Giant fans no longer called him a bad sport or a dirty player. He was spunky little Eddie. He was clever little Eddie. He was resourceful little Eddie. That was one of my earliest lessons in the relativity of ethics.

I told my questioner about this experience. It may seem frivolous to compare such an incident to the moral dilemmas of engineers, but I think the analogy is worth considering. Is it not true, at work or play, that we instinctively root for the home team? Of course, we do not condone real knavery; but we tend to view our own cause in a favorable light. An engineer who works for Exxon feels differently about offshore drilling than he would if he were employed as town engineer for a seaside resort. This has less to do with good and evil than it does with where one stands at a given moment. The petroleum engineer feels that he is called upon, both by his company and society as a whole, to provide an adequate supply of oil at a reasonable price. The town engineer is responsible for the well-being of the people who rely upon him to keep their community safe, clean, and prosperous.

Obviously an engineer's attitudes are in some measure formed by his assignment. Not every project need be exactly to his taste (as I argued in Chapter 7), but on the job he will likely find his fellow workers to be, by and large, decent people with whom he will want to join in common cause. From the outside, Exxon may appear to be a giant, heartless corporation obsessed with making profits. From the inside, Exxon doubtless is perceived as an association of earnest individuals working hard on constructive projects. "Groupthink" is a well-documented phenomenon from which none of us is immune.

"But isn't this bad?" the troubled student asked. "Not necessarily," I replied. Nobody wants a professional to pledge blind allegiance to an employer or a client, but since the *sine qua non* of accomplishment is effort, society is clearly well served if most people approach their work with zest—even if the zest is founded in something as apparently trivial as team spirit.

The student and I did not have time to pursue the discussion further. Later, however, I wondered if my answer had not rung hollow. After all, the classic response in such an encounter has always been "To thine own self be true." Yet I could not persuade myself that this was what one ought to stress to a young person preparing for a first engineering job.

In addition to enthusiasm, the good work that assures communal well-being requires order and cooperation. Any group of people undertaking a complex task must establish an organizational structure. Without assigned responsibility—without discipline—large-scale technological enterprise is unthinkable. Accordingly, a young engineer will have to temper any personal misgivings, such as they are, with an awareness of the group's needs for order. This may sound slightly unsavory, but only until it is considered in depth. Our daily lives teach us to live with disappointment. Our particular point of view cannot always, or even usually, prevail. In politics we learn to accept this and proudly call it democracy. We do what we can to influence the tide of events without being excessively disruptive. It is called working within the system.

Indeed, the system seems to work best when different groups, each energetically pursuing its own goals, clash and are obliged to resolve their differences. The term "healthy competition" has become a cliché, but it expresses a profound truth. Competition *is* healthy—ecologically, politically, and, of course, technologically. Our society's strength stems in no small measure from its multiplicity of often conflicting institutions: corporations, labor unions, bureaucracies, universities, trade associations, charities, foundations, political parties, and public interest organizations (which often more accurately could be termed "private interest organizations"). In this respect the body social reminds me of the human body, in which health is maintained by a variety of defense mechanisms: antiseptic fluids, white blood cells, the lymphatic system, and an incredibly complex chemical immune system.

Society's interests are better served, I believe, by the resolution of conflict between organizations than by the disaffection of individuals within organizations. Let oil companies search for oil—not recklessly, without care and common sense—but enthusiastically, without the kind of inner dissension that results in paralysis. Let the rest of us establish the limits beyond which we do not want the oil companies to go. And let the same process apply wherever technological progress impinges on other values that society holds dear—particularly public safety and environmental quality.

When engineers are loyal, each to his own organization, the system works and the public is served. But more is involved than efficiency. The quality of communal life depends upon the trust and respect that prevails among families, friends, and coworkers. An abstract devotion to "the good of humanity" is no substitute for devotion to real human beings. Out of fruitful personal relationships comes the decency that sets the moral tone for the good society. A sociologist who interviewed hundreds of workers once told me that what Americans love best about their jobs is the social milieu in which they spend their days—the relationships they establish with their colleagues. The Japanese have developed a sense of group harmony far beyond what we Westerners can, or care to, cultivate. But even in our individualistic society, mutual loyalty and good faith in the workplace is a key to personal happiness, and productivity as well.

Unfortunately, since the 1950s the concept of loyalty has been tarnished by too many loyalty oaths and abusive demands for conformism, just as the concept of patriotism has been sullied to the point where the term is on the verge of becoming pejorative. The problem is as old as humanity: how to balance the rights of the individual with the needs of the group.

I do not advocate unthinking obedience to the group, obedience of a sort that was confronted and discredited at the Nuremberg trial of war criminals. And I do not say that whistle-blowing is never justified. It is, occasionally, not only justified but necessary and heroic. However, one cannot endorse—except in the most exceptional circumstances—the betrayal of one's companions for some "greater good." Ayatollah Ruhollah Khomeini urges students to spy on their teachers and classmates and to secretly tell security forces about those whose dedication to Islamic values is in doubt. This is whistle-blowing of the worst kind, the other end of the spectrum from Nazi-type loyalty. Between the distasteful extremes of betrayal and servility each individual must make his own way. Happily, the need for engineers to make such choices is, to my best knowledge, exceedingly rare.

If I were able to resume my discussion with the troubled young engineer, I would urge him to embark upon his career with great enthusiasm and not be overly apprehensive that his work might harm society. Naturally, he must follow his own star, but to a certain extent he has done this already by choosing to be an engineer, a professional who must work with others, rather than, for example, choosing to be an artist, who can work alone. There are still personal decisions to be made. If he is averse to armaments or nuclear power, by all means let him steer clear of these fields. If he is an ardent environmentalist, let him seek work in that area. But once engaged in an engineering task, let him put his whole heart into it. He can maintain his values and sense of self without indulging his ego at every turn. If he wants to serve society, let him do good work.

What if, in spite of the enormous odds against it, he does encounter base practices and experiences a crisis of conscience? Loyalty to the group will require that he come down on the side of legality and prudence. No assemblage of engineers is well served by deception. And if the very worst should happen—if he becomes involved in a situation that cannot be honorably resolved within his organization—then, speaking as a member of society, I would rather count on the righteous wrath of an engineer whose loyalty has been betrayed than on the pique of an engineer who was from the start a suspicious malcontent.

It is a bittersweet paradox that new-fangled technology, because it depends so much upon group effort, should summon forth old-fashioned morality. The American philosopher Josiah Royce, writing at the beginning of this century, spoke of the need for "loyalty to loyalty."[1] The idea is worth reviving in our time.

10

Some Thoughts About Income—Monetary and Psychic

Engineering is a profession, and this fact moves us to explore its ethical and humanistic aspects. But, like all professions, it is at some times and in some circumstances a business—it is done for profit. What can we say about the relationship of profit to engineering, about the role of the "profit motive" in the practice of this learned profession?

This question came to mind repeatedly and with particular intensity a few years ago when I traveled to the Soviet Union, that strange and forbidding land where almost all technological work is performed in the name of the state. When Lincoln Steffens visited revolutionary Russia in 1919, he was mightily impressed. "I have been over into the future," he said to a friend, "and it works." Whatever else can be said about the Soviet system today, it does not appear to be working, at least not very well. One can refer to the massive study undertaken by the Congressional Joint Economic Committee *(Soviet Economy in the 1980s: Problems and Prospects)* or read scholarly academic works (for example, Marshall I. Goldman's *USSR in Crisis: The Failure of an Economic System).* Or, as an ordinary tourist, one can simply look and see. In the wealthiest sections of the largest Soviet cities there are hardly any fruits or vegetables to be had. There are no lively cafés or pubs in evidence, nor any of the colorful window displays that are such a prominent feature of Western cityscapes. There are few cars in the streets, and most telling, very little merchandise in the shops. While walking through these grim metropolises, it is difficult to believe that one is actually in a so-called superpower nation.

To be sure, the almost total absence of consumer comforts is partly attributable to policies of the Soviet government: the emphasis on basic heavy industries and armaments. Undoubtedly there are other factors involved, such as geography, climate, history, psychology, and the like. It is obvious, however, that the Soviet industrial system is badly crippled by the lack of economic motivation. Where the entrepreneurial spirit cannot be rewarded, and where non-productive workers cannot be discharged, stagnation will set in. Everything we know about human nature tells us this must be so, and these assumptions are born out by the frustrated statements of Soviet leaders as well as by the studies of Western experts. The tourist garners additional evidence as he encounters workers who are sullen and inattentive.

In my own field, building construction, the lethargy is almost palpable in the air. I can spot a sick construction job from a quarter mile away, and I saw such sick jobs all over Moscow and Leningrad. Tall cranes tower over sprawling concrete frames, and somewhere, almost lost in the immensity of the structure, a few men putter away connecting pipes or laying masonry block. Obviously some critical materials have not arrived—perhaps the door frames or the windows. Or maybe some crucial crew is not available—the electricians, say, or the sheetmetal workers. In free societies the marketplace would not permit such waste of invested capital. Bankers want their construction loans repaid and the interest clock runs without stopping. Manufacturers avidly pursue orders and workers seek employment wherever it is to be found. Contractors have learned to equate success with timely completion, and they dread the ruin that stalks the dawdler.

In the Soviet Union, not only do buildings take too long to build, but many of them start to fall apart before they are finished. Spalled masonry and cracked concrete are as much a feature of Moscow streets as wall posters featuring red stars and pictures of Lenin. So widespread are the problems with post-war apartment buildings that even a stubbornly patriotic guide admits, "At first we concentrated on quantity; now we must concentrate on quality." But if nobody gains when quality is good, and nobody loses when it is bad, it is hard to see how concentration is going to help.

In the marketplace of free nations, profit calls the tune. From dishwashers and automobiles to radios and pocket calculators, the products that fill our sparkling stores are created by people thinking of gain. Each consumer's desire—or potential desire—becomes an energizing force. Into every interstice of demand there is a rush of hopeful designers, manufacturers, and salespeople. And whatever products do not sell at retail quickly end up at an auction of remainders. No five-year plan can hope to equal such a super-sensitive organic system.

At the heart of this system—pulsing to the rhythm of profit and loss—is creative engineering. Clever ideas increase earnings. Money-saving procedures are sought out and rewarded. Creativity begets profit, and profit elicits creativity.

Of course, this efficient system has its social costs. We pay a high price in the form of unemployment, bankruptcies, and anxiety—and engineers bear their full share of these afflictions. The Soviet commitment to equality and full employment, however ineffectual it may be in practice, responds to human cravings for justice and security. Any capitalistic society that does not address such elemental needs cannot be considered satisfactory, no matter how wealthy it may be. There is also a dark side to the profit motive, an unpleasant complication that we acknowledge in our condemnation of avarice. At what point does wholesome ambition become unwholesome greed?

Our uneasiness in the face of this moral issue is especially intense when we contemplate the professions. We recognize that doctors, lawyers, and engineers—even ministers and teachers—

like money, but we find this disquieting. When IEEE's *Spectrum* published an article entitled "How to Make it Big: Engineers as Entrepreneurs,"[1] I read it with interest, indeed fascination, but also with a feeling of guilt. Every so often a story comes out of Silicon Valley—another engineer has converted his brilliance into an instant fortune. We thrill to the news and yet in some recess of our mind we wonder. Is this what engineering is all about?

Even as we ask the question we know the answer. The quest for profit is not the essence of engineering. Indeed, there is ample evidence that most American engineers are motivated by other forces. The field of public works is populated by thousands of engineers who work for modest salaries and find satisfaction in accomplishment and service. In academe and government many of the most accomplished engineers seek rewards that are inherent in research, discovery, and problem-solving. Even in the private sector, most engineers do not equate professional fulfillment with monetary gain. Dollar income is one measure of worth and reward, an important one perhaps, but by no means the only one. All engineers enjoy aspects of their work that have precious little to do with money. Very few engineers become entrepreneurs, and in fact the history of engineering is partly a tale of conflict between engineers and entrepreneurs. William R. Hutton, a leading railroad engineer of the late nineteenth century, resigned a top-level administrative position in protest against the chicanery of his business colleagues. John B. Jervis, a founding member of the American Society of Civil Engineers, threatened to withdraw as chief engineer of the Croton Aqueduct when its sponsors sought to use a mortar he considered inferior. Government engineers have long waged battle against industry, from the meat packers and gas companies of the past to the plastics manufacturers and power plants of the present.

The dazzling accomplishments of free enterprise, however, give rise to the persuasive but mistaken notion that only profit-motivated endeavors will succeed. In the United States this argument predates the Russian Revolution. At the beginning of this century there was a movement to put public utilities into the hands of private enterprise, on the theory that governments were by nature ineffectual and corrupt, and that government enterprise was accordingly suspect. An editor of *Engineering News,* in 1906, felt moved to protest: "The old argument that municipal ownership was to be abhorred because it was a step towards socialism can no longer be used."[2] In the ensuing years, government agencies have successfully undertaken many great works. Few people today suggest that our highways and water systems be turned over to private interests.

With these thoughts in mind, I see that my initial reaction to the Soviet Union was somewhat simplistic. While it is obvious that consumers are ill-served by the dominance of government, it is equally obvious that there are some things the Soviet government does fairly well. The trains run on time. The subways are a showplace. And clearly there are competent people working in the fields of space and weaponry.

As a tourist, one's view is distorted. I was taken into ancient monasteries and czarist palaces but not permitted to see factories and shipping depots, not to mention laboratories and space facilities. I learned more about icons than about Soviet technology. I did gain entrance to the Technological Institute in Leningrad, thanks to a sleepy guard at the door, but once I made my presence known to the authorities, I was politely but earnestly urged to leave.

So I gained no special knowledge about what motivates Soviet engineers. From what I have read, the best of them are granted perquisites that are the equivalent of handsome salaries in open societies. I have met a few traveling on government business, and they seemed cheerful enough, though not about to share with Americans their inner thoughts on career satisfaction. I suppose, like most of us, in addition to material rewards, they strive for honors and reputation. And, like engineers in all places and at all times, they also must be motivated by the pure fascination of the task at hand.

One can even speculate that their work serves to divert them, to help them forget the harsh realities of living in a totalitarian society. I am reminded of an imprisoned engineer in Aleksandr Solzhenitzyn's novel *The First Circle* who becomes totally absorbed in the technical project assigned to him: "This capacity to devote himself wholly to his work, to forget about life, had been the basis of his engineering triumphs on the outside, and in prison it helped him bear his misfortune."

There is a dark fascination in speculating about the Soviet Union, but in the end that enigmatic culture defies the analysis of experts, to say nothing of presumptuous tourists. As for the relationship between engineering and profit, that too resists precise definition. Profit is a spice, a spark, a catalyst—a crucial element of the technology that has flourished in capitalist societies and remade the world. The empty shelves of Soviet stores bear witness to the importance of the freedom to seek personal gain. But the achievements of engineers in socialist societies show that technology comes in great measure from creative wellsprings that have nothing to do with profit and recognize no political borders.

Speaking of profit, should engineers become managers in industry? The question invariably evokes feelings of ambivalence. According to one historical tradition, management is an integral part of technological creativity. For the fabled "chief engineers" of pyramids, cathedrals, bridges, and railways, design and construction were joined in a single organic process. In the shop also —whether the product was a steel ingot, a steam engine, or an electric transformer—conception and execution went together. In the early days of corporate industrialism, engineers saw management as an important aspect of their profession. Factory organization—even worker efficiency and motivation—were deemed accessible to technical analysis, giving rise to scientific management and industrial engineering. As I noted earlier, the author of a landmark 1929 report on engineering education was pleased to report that two-thirds of all American engineers embarked on "a healthy progression through technical work toward the responsibilities of management."[3]

On the other hand, as engineering has become more scientific and specialized, and as business management has evolved as an independent discipline, complete with its own schools and trappings, a schism has developed. Many engineers have recoiled from the managerial role and sought to maintain in technical design what they perceive to be professional integrity. The characteristics of a good manager—a feeling for people, politics, and the bottom line—appear to conflict with the characteristics of a first-rate creative engineer—a feeling for numbers, theorems, materials, and spatial relationships.

One's first instinct is simply to let each person find the niche to which he is drawn by his own peculiar talents and interests. Difficulties arise, however, when it turns out that managers earn more money, power, and prestige than their purely technical colleagues. The problem solves itself when engineers move willingly into management as they get older, but among those who wish to stay with technical work the result is often a festering resentment. They see rewards going to people who they consider to be less talented and who they believe, in the overall scheme of things, are making lesser contributions. For many organizations, both private and public, ill feelings among technical personnel create irksome problems of morale. In addition, when technically creative people move into management in pursuit of higher salaries and recognition, the loss of pure engineering talent is sometimes keenly felt.

One possible solution to this problem, instituted in some corporations after World War II, is the so-called "dual ladder." Where this system is in place, an effort is made to allow engineers to climb to the highest seniority levels by doing technical work, keeping pace with managers in a carefully planned schedule of promotions. However, according to a panel that studied engineering employment under the auspices of the National Research Council, this concept has not generally fulfilled its initial promise.[4] It works well only in large organizations and for limited numbers of engineers. A vice president of Honeywell's Corporate Technology Center has been quoted as being pleased with the dual ladder at his company while noting that in many organizations "the dual ladder comes and goes." "The dual ladder tends to flounder from an imbalance of power," he concludes. "The real power is in management."[5]

On reflection we can see that any attempt to correlate income and prestige with "worthiness" is doomed to fail. There is a certain rough-and-ready democracy in the marketplace, but it does not correspond to the personal values of most citizens. It is no secret that teachers, judges, and nurses are underpaid while basketball players, negligence lawyers, and stockbrokers receive more than they "deserve." Nor is unfairness limited to our own relatively free-wheeling society. In other countries, wealth and perquisites tend to gravitate toward commissars, generals, and black marketeers. Karl Marx said, "From each according to his abilities, to each according to his needs," but a catchy phrase does not signify a workable system. We look in vain if we seek a utopia in which the distribution of tangible rewards is totally just. It is against this broad panorama of human civilization that the problems of the technically creative engineer must be viewed.

There is a free market in job satisfaction as well as in dollar income. I am a civil engineer in the construction industry, and as such I have been a manager almost since the day I graduated from engineering school. I have had my share of compensations, but I note that when I and those like me meet with engineers who are instead engaged in design, we cannot help feeling traces of envy. Like the blondes in old hair-rinse commercials, *designers have more fun.*

On occasion, the technically creative engineer gets it all—wealth and fame along with the satisfactions of invention, as witness the tycoons of Silicon Valley. And at certain times and in certain organizations, engineers are given a full share of fortune and glory for performing technical work. For most engineers, however, the move from scientific craft to engineering management must be viewed as a natural process, like aging, that has its rewards as well as its drawbacks.

There will always be those who resist such a course, and I consider this a plus for society, the profession, and the human spirit. Still, I cannot see how these people can be satisfied in every respect. The dual ladder is a concept, imperfect at best, not a constitutional right. Happily, as Emerson said, "The reward of a thing well done is to have done it."

11

Educating the Public: Biting Off More Than We Can Chew

If the ultimate decisions about the use of technology are to be made in the political arena, then it follows that the public should know as much about technology as possible. I accept this as a given and believe, as I have already noted, that engineers are morally obligated to participate in the educative process. Yet, like most well-meaning ideas about the shaping of technology, this one ignores the complexities inherent in its application. The superficiality of the concept was made clear to me as a result of an experience that began in my dentist's office.

I had been informed that a cracked silver filling needed to be removed, and my dentist suggested that it be replaced with a gold inlay. I asked why the more costly material was to be used and was given various structural reasons that were more or less convincing. I agreed to the recommendation, and was about to drift into the trance that I try to attain in such situations, when the dentist added something that made me suddenly alert. "Besides," he said, "these so-called silver fillings really are amalgams that contain a lot of mercury, and we're beginning to wonder about the possibility of this mercury leaching into our patients' systems. I've been following the work of experimenters in Colorado, and pending further results I'm being a little cautious about using the stuff." When I told him that I had never heard of any such danger, he replied that it was not exactly the sort of information that the dental profession would want bandied about. "The research is in its very early stages," he said, "and we don't want to frighten people unduly."

As he proceeded with his work, my mind started to race. A potential catastrophe! Tens of millions of people—one of them me—with insidious mercury leaching from their teeth into their vital organs. And what a scandal! Research being performed and the findings not publicized. By the time I left the dentist's office, my alarm had subsided; but I determined, nevertheless, to find out more about the subject.

My research consisted of obtaining a newsletter published by the Toxic Element Research Foundation (TERF) of Colorado Springs. Its single urgent message was that toxic elements have no place in the mouth. The four-page leaflet seemed to be addressed mainly to dentists, but there was nothing in it that was technically obscure. Unfortunately, I could not tell whether it was the product of discerning professionals or the fantasy of eccentrics. I suspected the latter, particularly since a section captioned "Strong Testimony" consisted of an "unsolicited expression of appreciation" from a young man whose arthritic symptoms disappeared one week after two small amalgams were removed from his teeth. Yet the board members of TERF included a D.D.S., an M.D., a Ph.D. in biochemistry, a psychologist, and the president of a bank. I was feeling reasonably healthy at the moment, so I decided to forget about the whole thing.

I could have pursued the matter, of course, but then how would I have found time to keep abreast of the latest information on acid rain, the greenhouse effect, recombinant DNA, nuclear waste, new drugs, new diets, and a myriad other technical matters that affect me and the society around me? It is all very well to speak of an informed public, but in truth we spend our lives relying upon experts.

This is not necessarily as distressing as it sounds. Lewis Thomas, writing about a newly implanted pacemaker, reported with suprise and amiable guilt that the theories he had held as physician-philosopher changed completely once he became a patient. "Don't explain it to me," he said. "Go ahead and fix it."[1] A lot of people seem to feel the same way.

The conventional wisdom about a technologically informed public was challenged effectively a few years ago by Leon Trachtman of Purdue University. After a quarter-century devoted to studying, teaching, and writing about science, Professor Trachtman (in the quarterly *Science, Technology & Human Values)* questioned what he had come to think of as a "glib assumption." Efforts to inform the public about science and technology were rarely effective, he concluded, either in improving individual consumer choices or communal policy decisions. "When there is a scientific consensus," wrote Trachtman, "there is no need to inform the public except to recommend a proper course of action. When there is no consensus, why innundate the public with ambiguous and contradictory reports . . .?" Furthermore, "Since the important issues are generally the ambiguous ones, more knowledge seems almost calculated to create greater uncertainty."[2]

The argument impressed me very much, and brought me to the verge of changing my mind. But when I spoke about it to a few people I respected, they all warned me not to adopt an elitist position.

Shortly after the visit to my dentist, an issue of *The American Scholar* arrived in the mail and in it an article by Jeremy Bernstein entitled "Science Education for the Non-Scientist." Here it was again, the warning that unless we learn science, technological decisions will be made on our behalf—and to our regret—by others. (Of course, as Bernstein pointed out, there are other reasons to study science—not the least of which is pleasure.) Bernstein expressed optimism about the ability of a scientifically literate public to make its own decisions; yet what did he offer to support his conviction? He told of teaching a course to fourteen science majors at Princeton and bringing in four experts in nuclear power (both pro and con) to address the class. By the end of the term, the class had been converted from unthinking opponents of nuclear power to grudging advocates of nuclear for at least a partial solution to the energy problem. An interesting object lesson. And if science education for the masses could be based on this model, what a fine world it would be. But the trouble is that nuclear power is only one of hundreds of complex technological issues, and none of us can spend a semester on each.

The obvious problem with letting the few make decisions for the many is that the many, if they knew more, might want to do something different. There is a further concern, however. Some people fear the coming of "a new kind of Dark Age—a time when small cadres of specialists will control knowledge and thus control the decision-making process." (The quote is from *Higher Learning in the Nation's Service,* cited approvingly on *The New York Times* op-ed page by the president of Cornell University.)

I do not fear the coming of a sinister technocratic cabal, mainly because on consequential issues the technicians invariably give conflicting advice, and the politicians end up making the decisions whether they want to or not. (Jimmy Carter complained about this in the final days of his presidency, and he was surely not the first, nor will he be the last, elected official to do so.) Still, it would be nice to think that people could themselves make choices that intimately affect their own lives.

The paradox defies resolution. A wide diffusion of knowledge is good, but the uncertainties are awesome, no citizen can be adequately informed, and perplexing technical reports lead to harmful anxiety and erratic political action. As the saying goes, a little knowledge is a dangerous thing.

Ah, but this is not the saying, at least not as Alexander Pope coined it. Pope spoke not of knowledge, but of *learning;*

> *A little learning is a dangerous thing:*
> *Drink deep, or taste not the Pierian spring:*

There shallow draughts intoxicate the brain,
And drinking largely sobers us again.

Perhaps the semantic difference between *learning* and *knowledge* can give us a helpful clue. Learning is whatever can be acquired by systematic study, and a little of that—a few undigested facts—can indeed "intoxicate the brain." Knowledge in the broad sense, however, implies understanding, discernment, and judgment, and no amount of this, however small, can be a dangerous thing. A knowledgeable public will not expect to resolve each technical issue by analyzing evidence, but will seek to establish a fruitful relationship with its experts—and its politicians—a combination of trust and suspicion, respect and obstinance, calculated to best translate social objectives into technical decisions. A frank appraisal of how each individual reaches decisions about technology reveals that it is largely a matter of deciding which experts are worthy of one's confidence. This does not diminish the need for wide-spread education in science and engineering, but it does indicate the superficiality of many current pronouncements on the topic. Educators and science journalists will have to seek new ways to responsibly assist in the complex process of decision-making.

I believe that we have been too simplistic in our assumptions about science education and the public. While I'm rethinking the big picture, however, and resigning myself to reliance upon experts, I plan to keep asking my dentist what they're discovering—if anything—about my fillings.

12

The Fall of Rome—According to the Press

A CLUE TO THE DECLINE OF ROME.

The *New York Times* headline captured my interest, as the paper's editors knew it would. The decline and fall of the Roman Empire is a topic of perpetual fascination, and it raises further questions about technology and the public. One's natural curiosity about Rome is heightened by the uneasy awareness that our own society is showing signs of decay.

The headline was occasioned by newly discovered evidence that the ancient Romans suffered from lead poisoning. Dr. Sara C. Bisel, an archeologist and physical anthropologist, examined the bones of fifty-five victims of the Mt. Vesuvius eruption of A.D. 79 and reported her findings at a meeting of the American Association for the Advancement of Science. Dr. Bisel found that the bones contained a mean level of eighty-four parts of lead per million, which, she concluded, was unusually high. (The bones of typical contemporary Americans and Britons contain twenty to fifty parts per million; in the remains of prehistoric Greeks, the figure has been as low as three parts per million.)

It has long been known that the Romans used lead for water pipes and kitchen utensils, and that much of their food and drink was contaminated. A few historians have suggested that this so seriously affected the health of the citizenry that it inevitably brought about the collapse of the empire. Because Romans practiced cremation of the dead, not many bones have been available for analysis. But, according to the *Times* article, the recent discovery of a large number of skeletons at Herculaneum, near Pompeii, provided the opportunity for the new study. The mention of Herculaneum and Pompeii piqued my interest even more, since I had visited those famous archeological sites just a few years earlier. At that time, it so happened, I had my own experience relating to lead and fallen empire.

The ruins are truly spectacular, bespeaking triumphs of ancient engineering, and they stirred in me strong feelings of professional pride. Most of my co-travelers, however, were nontechnical types who seemed more interested in myths and statues than lintels and paving stones. (It is only fair to report that my wife and I were on a trip sponsored by a "Hellenic" association of London.) Our leaders were classical scholars of great charm and erudition, but I thought their praise of ancient engineers excessively muted. "A nice piece of work," one of them admitted to me as we stood in front of a tower that even today would challenge the skill of a master builder, "but, after all, they had all those slaves!"

At site after site, marvels of engineering were ignored in favor of architectural niceties, until at Pompeii we came upon lead water pipes, and then technology suddenly took center stage. Everyone started to talk about the danger to an unsuspecting populace that resulted from the use of the hazardous material. "Always have to try something different, you engineer chaps," said one of my companions. "Tried to poison everybody in town," said another, chuckling, "just like today." It was all lighthearted raillery, but I felt as if my good nature was being tested to its limit.

It is bad enough to ignore the vital contribution of engineering in the growth of classical culture and to speak as if the most significant feature of a great stone column is not the miracle of its quarrying, transport, and erection, but merely the carved decoration at its top. Having done this, then to imply that technological advance brought about the *downfall* of ancient civilization —well, that is simply too much!

This feeling of exasperation came rushing back to me as I read the *Times* article. The attribution of Roman decline to widespread lead poisoning has always seemed to me an untenable theory. The symptoms of the disease are the appearance of a blue line on the gums, weakness, anemia, colic, and paralysis of the wrists and ankles. In extreme cases, after massive exposure, there may be brain damage and death. If this condition was widespread in the empire one would think that the written record would be replete with examples, which it is not. Lead doubtless caused health problems, but there is no indication that the empire collapsed because of lead-induced illness or indeed any particular disease. If a health related lassitude is thought to be the cause of decline, then I find somewhat more convincing the theory that malaria was the culprit. But this is historical nit-picking.

The great mystery about the Roman Empire is not why it declined, but how it lasted as long as it did. Rome collapsed because it grew old and could not cope with social change. Today we are grappling with the same problem: how to convert from a pioneer society to a mature one, or better still, how to maintain the vigor of the pioneer and apply it to a new array of challenges.

Just as Rome did not collapse without being assaulted from outside the Empire, so is America's decline tied to the rising demands of other societies. The problem in Roman times, as it is now, was never too much technology but rather too little to go around. If the tribes of the Asian steppes had been warm and well fed, they would not have invaded Europe. If, today, other societies were not competing with us over scarce resources and costly energy, we would certainly feel less threatened.

Historians tell us how the barbarian tribes learned and adopted many of the customs of the Romans. Our "barbarians" —and here the word is used in its original sense meaning "foreigners"—have done the same. We, however, are doing something that the Romans did not. That is, we are studying our challengers just as they have studied us, and bid fair to adopt some of their techniques for ourselves. I refer not only to Japanese management systems, but also to Oriental methods of controlling personal stress, Third World alternative technologies (such as small methane generators), and sophisticated European trade practices. Our task is made somewhat easier in that we are not trying to stay at the very top of the world, as the Romans were—we just want to maintain the standards we have achieved and protect our opportunity to improve them. We see the salvation of the world not in the exploitation of others, but in the improvement of conditions for all. There are a lot of things working for us that the Romans did not have: technologies beyond the wildest imaginings of the engineers of antiquity, statistics and theories from the social sciences, a tradition of justice more appealing than Rome's, and, not least, some knowledge of history.

If we divert ourselves with false historical theories, we reduce our capacity to address the problems at hand. If we think that Rome fell because of its plumbing and cookware, we will be less apt to think in terms of world commerce and global understanding.

There is, to be sure, a valuable lesson in the *Times* article. It relates not to empires but to the nevertheless very important field of public health. Repeatedly throughout history, technical innovations have had adverse consequences that were not anticipated. When pumps were developed, for example, the need for aqueducts diminished, as drinking water was obtained from rivers and wells near or in cities. As a result, contaminated water caused diseases that had not been a problem when the water came pure from distant mountains. Such predicaments bedevil us to this very day, and it is instructive to consider case histories wherever they may be found—past or present. From a medical point of view, it is interesting to speculate, as some experts have, about the extent to which the eccentric behavior of some Roman emperors might have been related to the lead content of their wine. But when one considers the way these demi-monsters spent their childhoods, psychological theory is adequate to explain almost any aberration.

I find no fault with the scientists who study the phenomenon of lead in Rome, nor with the science writers who report their findings and their theories. But when a cautionary tale about public health is transmuted into a fairy tale about history, I begin to fear that we are losing our sense of reality. Our civilization will survive only if we address ourselves successfully to the great problems of economic development, social justice, and political equilibrium. As for environmental pollution—and nuclear proliferation, that other "technological" source of public anxiety—engineering solutions are at hand, but political approaches are not. I hope that historians will not look back at us and find that we refused to face up to the challenges of our time, and that we fell into the habit of blaming technology for the very decline it might have helped to avert.

13

The Fantasy of the Electronic Future

For every article that equates technology with the fall of civilizations there must be at least two that portray it at the opposite extreme of human potential—as a magic carpet to Utopia. Much as I enjoy reading about engineering marvels, I sometimes think that effusive optimism is even more misleading—and more damaging in the long run—than doleful pessimism.

Several years ago the editors of *Omni* magazine celebrated their publication's first anniversary by inviting a number of noted scientists, engineers, and science writers to speculate about the future, and the resulting special issue was a striking example of the buoyant overconfidence that irritates me no end. Isaac Asimov, for example, predicted that microcomputers would make it possible to proceed with the total automation of society, "therewith removing the necessity for humans to indulge in dull and repetitious labor." Ray Bradbury opined that space travel was "commensurate with the birth of Jesus, Buddha, or Muhammad." René Dubos said that he saw no limits to "advances in technology and ways of thinking." Various contributors spoke excitedly of solar energy, recombinant DNA, ultra-intelligent machines, the homeostatic handling of pain, and undreamed of breakthroughs in the physics of elementary particles.

These visions made for stimulating reading. But as I sat in my armchair one evening, turning the glossy pages of the magazine, I felt as if I was drifting into a world of fantasy. This was the future discerned, all right, but it seemed somehow ethereal.

Hopefulness is nice, particularly on an anniversary, but something important was totally absent from these visions, something I couldn't quite put my finger on.

The next day, as I drove my car on New York City's West Side Highway, rattling over familiar bumps and crevices, I abruptly realized what was missing from *Omni*'s picture of the future. It is all very well to speak of computers, lasers, and genetic engineering, but every day the actual road to the future unfolds before us and anyone with eyes can see that it is full of potholes.

The highway on which I was driving, built when I was a boy, had become a shambles of cracked concrete and rusting steel. Under New York's patched streets, ancient water mains leak and occasionally break, loosing great geysers on the long-suffering public. The docks are falling apart and the subways are a disaster of mythic proportions. In aging buildings, roofs deteriorate, windows rot, and worn cornices come loose, sometimes with lethal results. Seen from the vantage point of a rattling automobile, the great scientific and technological challenge of the future has little to do with electronics, DNA, or space travel. Essentially, it is how to renew and maintain the engineering works that were put in place during the past century.

I soon discovered that this insight was shared by many people in many places. The ink was hardly dry on *Omni*'s first anniversary issue when there seemed to sweep across the nation a wave of complaints and warnings about dilapidated public facilities. On all sides one started to hear the phrase "deteriorating infrastructure."

One focus of the widespread malaise was *America in Ruins,* a report issued in 1981 by the Council of State Planning Agencies. This study attracted media attention and inspired a feature article in *Time* magazine entitled "Time to Repair and Restore." The facts were widely disseminated and discussed in a variety of forums. The nation's 42,500-mile Interstate Highway System is deteriorating at the rate of 2,000 miles per year, and more than 8,000 miles require total rebuilding. One of every five bridges requires major rehabilitation or total reconstruction. Half of our communities must cope with marginal water supplies and a third with inadequate sewage treatment facilities. Almost half of Conrail's lines in the Northeast will probably have to be abandoned in the near future. Ports, prisons, dams—almost all public facilities—are in a deplorable state of decay. Since 1965, the percentage of the United States' gross national product invested annually in public works has fallen 52.8 percent, from 3.6 percent to 1.7 percent. Federal aid programs are being cut and the municipal credit market is in disarray.

According to the Council of State Planning Agencies, this fast-developing crisis escaped broad attention partly because the federal government does not have a separate capital budget. Thus the decline in capital investment is obscured by the rise in annual operating expenses. The Comptroller General of the United States also stressed this point in a report issued in 1981. The Office of Management and Budget, however, has resisted changes in this procedure, maintaining that if capital investment were to become a separate policy area, pressure politics would inevitably force the federal budget upward to unacceptable levels. It may be true that federal budgeting procedures contribute to the growing crisis, but it is equally true that states and cities, practically all of which *do* isolate capital expenditures, have neglected essential maintenance and renewal programs because of political timidity in the face of taxpayer revolts.

Having neither glamour nor overt urgency, the problem will not be solved quickly. The National Council on Public Works Improvement, established by Congress in 1984 and charged with recommending new policies prior to 1988, had barely gotten started before its operating budget was reduced by 75 percent.

It is ironic that while public facilities are literally crumbling about us, the future of science and technology should still be seen in terms of glittering novelty. This view would be understandable if our new techniques made it possible to forget our earlier creations, as when we built railroads and simply abandoned most of our canals. But this sort of thing is not about to happen again, much as we like to talk about sitting at home and communicating electronically. So far, people show no signs of being satisfied with seeing each other only on glowing screens. Not only do people want to move about physically, but the millions of tons of materials that form the basis of our civilization must still be transported by land, sea, and air—from fields and mines to factories to distribution centers to consumers. One of the key arguments of those who decry present government policy is that future economic expansion will be inhibited by a deteriorated transportation network.

The situation is made worse by the fact that new technologies create unforeseen maintenance problems almost from the moment they are introduced. Plastic water lines leak unexpectedly, foamed insulation proves injurious to the occupants of houses, newly created chemicals contaminate water supplies, offshore structures experience mysterious fatigue stresses, polymer gaskets in nuclear plants undergo unanticipated degradation when exposed to low-level radiation, and so forth. This is all very annoying, particularly to a society that is used to moving forward with its eyes on the stars.

However, such difficulties have their positive aspects. If into the foreseeable future we will be patching pavements and replacing gaskets, then robots are not about to take over and the age of the trowel and the wrench has not come to an end. Concerns about technological unemployment will prove to have been overstated. In science and engineering a host of new challenges are seen to exist. Just as doctors are beginning to rethink their attitudes about geriatric medicine, so will scientists and engineers see new opportunities in solving problems of preservation and renewal. This is already happening, although one would never think so by reading *Omni*.

* * *

A good place to look for a different view is the monthly journal of the American Public Works Association. Here one can read about improving landfill techniques, rehabilitation of maintenance equipment, renovation of an old waste water treatment plant (including new computerized controls), and the recycling of concrete pavement (by fracturing it in place and applying a bituminous surface). Evidence of the new reality is beginning to accumulate in many places. The Interior Department's Office of Surface Mining announces a television-guided method of hydrologic back-filling of abandoned coal mines to arrest the subsidence of Pennsylvania towns. Chemical companies are hard at work developing epoxies, paints, and other substances that will patch, clean, cover, and generally preserve existing installations. Bridges are being resurfaced with polymer concrete, deteriorated sewers relined with polyurethane, old roofs surveyed from the air with infrared scanners. Recently the American Consulting Engineer Council awarded its grand prize for engineering excellence to a firm that purified ground water contaminated by vinylidene chloride and phenol. The project entailed pumping the water from purge wells into aeration ponds, then using carbon adsorbers, and finally utilizing spray irrigation into a marshy area. Other prizes were awarded for the rehabilitation of New York City's Manhattan Bridge, the replacement of a railroad bridge in Chicago, a dynamic stability analysis for existing earth fill dams in earthquake areas, the cleanup and containment of PCBs around a General Electric Plant in Oakland, California, and a new bus maintenance facility for Houston.

Even lowly potholes are coming in for their share of attention, as well they should considering that each year in the United States approximately two hundred million of them have to be filled. The problem has been studied at the U.S. Army Corps of Engineers Cold Regions Research and Engineering Laboratory in Hanover, New Hampshire. One finding is that, in the long run, holes simply filled and compacted by hand cost five times as much as holes that are trimmed, cleaned, dried, lined with a tacking material, then filled and finally compacted by machine. This is not exactly the sort of discovery for which Nobel Prizes are awarded, but it is nevertheless something worth knowing.

In facing up to the infrastructure crisis we will have to temper our futuristic zeal with a stolid sense of reality. This does not mean succumbing to contemporary despair, but rather acknowledging a certain wisdom of the ages.

The lessons we are learning remind me of a painting by Pieter Bruegel the Elder that I saw several years ago in Vienna. Entitled "The Tower of Babel," this famous work shows a mammoth circular building being constructed, its top just touching the clouds. The architectural detail is magnificent and the depiction of sixteenth-century building techniques is exquisitely rendered. Obviously, the tower has been under construction for many years, probably generations. At the highest levels the new stone walls glow orange against the pale sky. The foundations, however, are dark and beginning to crumble.

14

Down to the Sea in Print

Newspapers and popular magazines are not the only places where one is liable to get a distorted view of technology. Within the engineering profession itself the electronic revolution has tended to obscure the more down-to-earth aspects of technical enterprise.

Like most engineers, I receive in the mail many unsolicited bulletins and announcements. The folders, cards, and booklets come in a variety of shapes and sizes, and they arrive in large numbers. I read this literature with an intensity that I find difficult to explain. I suppose that in browsing through schedules and scanning the titles of lectures I expect to learn something by osmosis. It is also comforting to know a bit about what is "happening" in the profession even if the details remain obscure. Whatever the reason, these mailings afford me a good deal of interesting reading.

Recently, however, this pastime has become less enjoyable. The change is related to the nature of the events that are being announced. The ubiquitous computer has taken over with a vengeance. Almost every gathering of technically trained people appears to be devoted to modeling, simulation, imaging, or optimization. Even in my own field, civil engineering, the emphasis is on the application of computers to drafting, estimating, and design. This is all well and good, except that engineers should know, better than anyone, that not all of our problems will yield to the application of software and microchips. Knowledge may be power, as Francis Bacon said, but power is only the potential to perform work; it is not work itself. An information explosion is not a real explosion; it does not blast rock for tunnels or foundations. The most brilliantly conceived computer will not raise crops, dispose of toxic waste, or repave highways.

Occasionally, to be sure, I receive notice of meetings devoted to large-scale engineering, but these have become relatively infrequent. They also seem to be less buoyant and full of promise than they used to be. Instead of new projects and new techniques, increasingly they deal with structural failures, the problems of government regulation, and the rising cost of malpractice insurance. Then there are the short courses given one recent summer by the Federal Emergency Management Agency. The featured offering dealt with fallout shelter analysis, hardly the most up-beat aspect of engineering.

And so it has been for quite a while now. Between the breezy promises of the electronic wizards and the often depressing problems inherent in macro-engineering, my pleasure in reading brochures about professional meetings has been greatly diminished.

There has been one striking exception, however. On a cold winter's day there arrived unbidden—and from what mailing list I cannot begin to guess—notice of an Offshore Technology Conference scheduled to be held in the Houston Astrodomain. I have no more experience with offshore technology than I do with electronics, but nevertheless the program gave me a lift. For one thing, the cover, instead of showing the pattern of an integrated circuit (a symbol of which I have grown exceedingly weary), featured a painting of the open sea, whitecaps breaking against an oil-drilling platform in the middle distance. Visions of the sea are enough to start the pulse racing in any person of feeling or imagination.

I do not usually think of oceans in terms of engineering, but obviously this reflects my own limited horizons. As I learned from this slim booklet, the Offshore Technology Conference has been held annually since 1969 and is currently sponsored by eleven professional societies representing more than five hundred thousand engineers.

The variety of subjects to be discussed at the conference was dazzling. One session would be devoted to the dynamics of offshore structures, others to foundations, platform construction, mooring and anchoring. Lectures were planned on seabed mapping and soil testing; seismic and sonar surveys; wire rope, anchors, chains, and hawsers; as well as erosion, abrasion, metal fatigue, and corrosion; deep-water drilling, welding, and tools for divers. Meetings were devoted to such disciplines as marine geology, arctic operations, the analysis of wave forces, and the study of ice. Most of the announced participants were affiliated with the oil industry, as one would expect. But one session was allotted to "Marine Minerals Mining" and the lecturers came from institutions I never knew existed: Minerals Management Service, Lockheed Advanced Marine Systems Division, and the University of Petroleum and Minerals, Dhahran. One presentation, however, was to be made by three gentlemen from an organization I certainly had heard of—Japan's Ministry of International Trade and Industry, the formidable MITI. Their topic was "Research and Development Project of Manganese Nodule Mining System in Japan," and it started me wondering about whether the United States might be falling behind in a technological field that is just about to come into its own.

The program for the Offshore Technology Conference was captivating, and I pored over it for a long time. It told me much about engineering activities to which I had previously given little thought. Of course, many brochures, catalogs, and menus promise splendors that in actuality turn out to be disappointing. Has the time for ocean engineering truly arrived, I wondered, or is the field—aside from offshore drilling for oil—still mostly dreams and hype?

I noticed that one of the sponsors of the conference was the

Marine Technology Society, another organization whose name was new to me. Upon investigation, I discovered that the Society was founded in 1963, has sixteen sections in the United States and Canada, fifty-seven institutional and industrial sponsors, and carries out its activities through four standing committees and twenty-seven professional panels. These panels are devoted to such subjects as Marine Food and Drug Resources, Marine Mineral Resources, Ocean Energy, and Coastal Zone Management. A typical issue of the society's journal will carry articles on a revolutionary ship design (featuring two torpedo-shaped hulls that are completely and continuously submerged), a laser system for ocean floor imaging, prospects for harvesting Antarctic krill (the biomass of these minute crustaceans exceeds the world's tonnage of fishery products), and the properties of electrodeposited minerals in seawater (the immediate aim is to protect and enhance the strength of seawater structures, but the prospect of obtaining minerals from seawater by cathodic electrodeposition is mind-boggling).

On March 10, 1984, the administrator of the National Oceanic and Atmospheric Administration announced the commencement of "The Year of the Ocean," an eighteen-month period in which the marine community intended to draw public awareness "to the value of our coasts and the ocean as a multiple resource for the nation." If there was any publicity I seem to have missed it. The March 10 date was selected to mark the anniversary of President Reagan's proclamation of an Exclusive Economic Zone extending two hundred miles into the ocean from the coasts of the United States. This proclamation followed in the wake of our nation's refusal, in 1982, to endorse the international Law of the Sea Treaty, a fact that clouds the prospects for future development of the oceans.

Cloudy prospects or not, investment in ocean engineering is a significant factor in the world's economy. The worldwide market for businesses serving ocean industries exceeds $200 billion.

Offshore oil accounts for much of this, but not as much as one might think. "Seventy percent of the dialogue," says a publisher of *Sea Technology,* "but considerably less of the money."

I am no soothsayer, least of all when it comes to incipient technologies. I tend to side with those who challenge the prophets: "If you're so smart, why ain't you rich?" Is nuclear power really on the skids? Is the breeder reactor never to be? Is fusion twenty years away, or a millenium? How about solar energy or synfuels? Is space the next frontier? Will the great advances come in biomedical engineering, materials, transportation, environmentalism, or mainly—as so many people seem to believe—in electronics? Or are we approaching the age of the oceans?

I do not know. But I do know that if I were starting my engineering career today I would be greatly tempted to look to the oceans. They are truly, literally, our next frontier. Their allure and the challenges they pose are unparalleled.

At least that is the way I felt after looking through the program with the seascape on the cover. Who knows what the mail may bring tomorrow?

15

Disasters and Decision-Making

In the best of all possible worlds, all engineers would be conscientious and all members of the public would be well informed about science and engineering. But even in such an ideal state it is unlikely that appropriate action would be taken to govern the course of technology. A missing element would still be, as it is so often today, that state of alertness—vigilance if you will—that underlies far-sighted planning. Without periodic shocks, people tend to become complacent, and complacency is the enemy of precaution.

When, in June of 1983, a section of Interstate Highway 95 plunged seventy feet into Connecticut's Mianus River, there was widespread relief that the accident killed only three people. If the bridge span had fallen during the day instead of at 1:30 A.M., the carnage could have been horrible. Newspaper stories painted gruesome pictures of what might have been. Within a few days, however, public concern subsided, and since this nonchalance was clearly related to the low death toll, perhaps the blessing must be considered mixed.

No right-thinking person will wish for casualties in order to make a point. (I recall the revulsion with which I read an article by an anti-nuclear advocate entitled "What This Country Needs Is a Meltdown.") But there is no denying that in the absence of outrage many things are ignored that ought not to be ignored, and nothing produces outrage as readily as large numbers of simultaneous, accidental civilian deaths.

From 1825 to 1830 there were forty-two recorded boiler explosions aboard American steamboats, each one killing an average of six to seven people. But when in 1830 more than fifty passengers were killed by one explosion aboard the *Helen McGregor* near Memphis, public indignation finally reached the point where Congress was obliged to take action. The result was the first technological research grant made by the federal government, followed in due course by regulatory legislation. Of course, the evolution of moral, social, and political values depends upon more than disasters. In mid-nineteenth-century America the idea of government responsibility for the safety of its citizens was just beginning to mature. But without attention-demanding events such as the *Helen McGregor* explosion, the translation of concept to practice can be unduly prolonged.

In our own time, the National Dam Safety Act of 1972 was passed after one hundred twenty-five lives were lost in the failure of a coal tailings dam at Buffalo Creek, West Virginia. As memory of the disaster faded, however, the inspection program mandated by the legislation was not funded. Inevitably, there was another catastrophe: in 1977, thirty-nine students and teachers at a Bible college died in a dam failure at Toccoa, Georgia. Only then did Congress vote funding for the program it had authorized five years earlier.

It is not only the number of dead that gives an event its shock value. In May 1979 a piece of masonry fell from the facade of a building in New York City and struck a young woman on the head, killing her. The building was owned by Columbia University, and the victim was a student at Barnard College, which adjoins the Columbia campus. For several days the event made headlines. Stories about the tragedy were followed by reports about dangers inherent in the city's aging buildings. Public concern led to demands for action, and early in 1980 the City Council enacted Local Law 10. The statute requires owners of large buildings to have the facades inspected every five years by a licensed architect or engineer, and to promptly repair any potentially unsafe conditions discovered. In this case, although only a single life was lost, the "dread" factor was extraordinarily high. The death of a student at an elite women's college—all that vitality and promise snuffed out—had news appeal. Also, every resident of the city walks its streets and wants to be protected from falling masonry. Another reason for the new law's popularity was the fact that it entailed no direct cost to the taxpayers. All expense, naturally, is borne by landlords.

Whatever the public chemistry of this event, it clearly led to action. In fact, as the repair program progressed, questions were raised about it being an overreaction. Hundreds of millions of dollars were allocated for the restoration work. Columbia University alone committed $17 million to the repair of its buildings. Real estate analysts warned that the additional expense might prove ruinous to some property owners. Aesthetes and preservationists complained as decorative friezes were hurriedly removed, often replaced with ugly patchwork stucco. Subjected to a cost-benefit analysis, Local Law 10 would surely fail. There is no logical basis for spending hundreds of millions of dollars to save perhaps one life every few years. Or, to put it more humanely, if we are interested in saving lives, there are far better ways of spending hundreds of millions of dollars. But we refuse to be intimidated by the mathematics of risk analysis. We simply will not tolerate buildings crumbling above our heads. And the death of that single Barnard student remained vivid in the communal imagination.

So New York City's buildings are being inspected and repaired, and the nation's dams as well, each program the result of dramatic accidents that captured, more than fleetingly, the attention of a fickle public.

In contrast, the collapse of the I-95 bridge was not an event of lasting traumatic impact. Although there were some negative comments about the span's design—a now obsolete method that relied upon a pin-and-hanger assembly—the basic cause of failure was found to lie with inadequate inspection and maintenance, and that does not make very good newspaper copy. Only three people were killed—a truck driver and two other men who nobody ever heard of. Public reaction centered around the traffic problems that developed while the road was out of service. It is true that the State of Connecticut resolved to improve its bridge inspection procedures, and across the nation there was a momentary flurry of concern. But the basic problem remained unchanged. According to the National Highway Traffic Safety Administration, half of the nation's 565,000 bridges are potentially unsafe. Indeed, about 150 bridges are washed away in floods each year and an equal number simply sag, buckle, or collapse. But these failures have occurred without loss of life—at least they have since 1967, when a span across the Ohio River between Point Pleasant, West Virginia, and Kanauga, Ohio, carried forty-six people to their deaths.

The issue of maintaining bridges is so mundane that our society seems incapable of attending to it. Yet this very mundaneness is what makes the neglect so disgraceful. There will always be engineering failures. But the worst kind of failures, the most inexcusable, are those that could readily be prevented if only people stayed alert and took reasonable precautions. The inspection, maintenance, and repair of bridges is neither glamorous nor newsworthy. Yet it is appalling to think that this important work might languish until a suitably shocking disaster occurs.

We need not accept it as a given that the public will always be deaf to reasoned warnings, responsive only to shock. Nevertheless, experience teaches us that society requires a cadre of concerned citizens—engineers foremost among them—to urge proper action and to persist when rebuffed. Where the community remains indifferent and politicians evasive, public-works engineers have no choice but to redouble their efforts, doing the best they can with curtailed budgets and inadequate staff.

Engineers, being human, are also susceptible to the drowsiness that comes in the absence of crisis. Perhaps one characteristic of a professional is the ability and willingness to stay alert while others doze. Engineering responsibility should not require the stimulation that comes in the wake of catastrophe.

Alertness, however—important as it may be—is only one component of a satisfactory technology policy. There still remains the question of how best, methodologically, to cope with technological issues.

16

The Deceptive Allure of Risk Analysis

Let us imagine that our dreams have materialized: engineers are conscientious, politicians are responsible, the public is well-informed, and everyone is alert and far-sighted. What procedures should we use to best achieve our technological objectives?

Much has been written in recent years about *risk analysis*—also known as risk-benefit analysis, cost-benefit analysis, risk assessment, risk comparison, risk accounting, cost-effectiveness analysis. "Regulatory action," proclaimed President Reagan shortly after taking office, "shall not be undertaken unless the potential benefits to society outweigh the potential costs." Thus risk assessment was placed at the center of national policy. At about the same time, Congressman Don Ritter of Pennsylvania, one of the two or three engineer members of the House, was sponsoring legislation that would require all government regulators to be guided by standardized, stringently scientific analysis of comparative risk.

In deciding to foster the cause of risk analysis, Mr. Ritter exhibited a political astuteness to match his professional credentials. (He earned a doctorate in metallurgy from M.I.T. and taught for ten years at Lehigh University.) At a time when the purported excesses of government regulation were the subject of intense interest and resentment, his proposal to subject regulations to scientific scrutiny quickly gained him a large and attentive audience. The hearings on his bill—which called upon the Office of Science and Technology Policy of the executive branch to "apply and promote" the use of comparative risk analysis by all regulatory agencies—drew overflow crowds for two days in May of 1980. "Mr. Speaker," intoned Congressman Ritter in calling attention to the bill on the floor of the House, "comparison of risks is clearly an idea whose legislative time has come. . . . Comparison of risks is a way for regulatory agencies to do a better job of protecting the public—by targeting their limited resources on the most serious threats to health and safety instead of regulating 'blindly.' Comparison of risks brings government regulation into the nineteen eighties."[1]

Hear, hear! I said to myself as I read these remarks. Yet the more I thought about this wondrous new analytic approach, the more uneasy I became. If there were truly reliable methods of analyzing and comparing risks, why would regulatory agencies not be using them? And if these methods were being resisted, perhaps they were not as wonderful as advertised. The few articles on the subject that I found in technical journals were bland and inconclusive. I decided that the best way to learn more would be to read the testimony of the people who participated in the hearings on Mr. Ritter's bill.[2]

The hearings took place before the Subcommittee on Science, Research and Technology of the Committee on Science and Technology, and they started out promisingly enough for Mr. Ritter, as several of his colleagues showed up to laud his new initiative. Representative William C. Wampler of Virginia expressed the hope that risk analysis would eliminate the "confusion and chaos" of current regulatory practice, particularly the unwarranted curbing of pesticides and food preservatives. Mike McCormack of Washington warned that quibbling over the negligible risks of nuclear power was "keeping this country from moving forward." Other representatives deplored the FDA's conservative approach to the approval of new drugs and its attempt to ban saccharin, the EPA's caution concerning diesel fuel emissions, and the Nuclear Regulatory Commission's timidity about releasing krypton gas into the atmosphere during the cleanup of Three Mile Island. It seemed as if the hearing had attracted every congressman with a grievance against a regulatory agency. Only Tom Harkin of Iowa, apologizing for being "a thorn amongst the roses," counseled caution. Mentioning the examples of thalidomide and Agent Orange, he suggested that if we make mistakes it is well that we err on the side of safety.

After the visiting representatives had chanted their prologue —and then quickly scattered to various other hearings and meetings—it was the turn of the invited technical experts. The first of these, Dr. James Vaupel of the Institute of Policy Sciences and Public Affairs at Duke University, sought to dramatize the benefits of risk analysis by evaluating the often-heard admonition against eating too many eggs. Although eggs are high in cholesterol, and in some individuals may contribute to atherosclerotic disease, the risk entailed in eating them has, according to Dr. Vaupel, been greatly overstated. By cutting in half the average American's consumption of eggs (five per week) it is estimated that five thousand lives might be saved each year. Although this sounds like something worth doing, when stated in different terms it can be shown to be practically insignificant. For example, this proposed halving of egg consumption will extend the average American's life expectancy by only ten days; thus each egg consumed over an average lifespan of seventy-three years might reduce life expectancy by about a minute; further calculation shows that eating a half-dozen eggs involves one chance in a million of a death per year, thus qualifying as a "micro-risk." Other micro-risks, for the sake of comparison, are: smoking 1.4 cigarettes, living two days in the air pollution of New York or Boston; traveling ten miles by bicycle or three-hundred miles by car; consuming Miami drinking water for one year; eating forty tablespoons of peanut butter (due to liver cancer caused by aflatoxin B). These are the sorts of comparisons that Rep. Ritter would require regulatory agencies to consider and put before the public.

Although the purpose of Dr. Vaupel's presentation was to support Mr. Ritter's bill—and, incidentally, to ridicule what he called "the Humpty-Dumpty assault on eggs"—its effect on me was to arouse a long-standing misgiving about the manipulation of statistics in public discourse.

The next witness was Dr. Walter Albers, head of the Societal Analysis Department of the General Motors Research Laboratories, another booster of the Ritter legislation, whose testimony considered the comparative cost-effectiveness of several life-extending programs. A coronary ambulance system, he noted, can save lives at a cost of $192 per person-year of longevity; for highway rescue helicopters the estimated cost is $4,180. By comparison, the cost of installing four headrests per car is $500,000 per person-year, and the cost per person-year of achieving EPA's objective of 3.4 grams per mile of carbon monoxide in auto emissions is $27,500,000.

When Rep. Ritter asked Dr. Albers how he felt the regulatory agencies would respond to guidelines requiring them to utilize risk analysis, Dr. Albers responded, "I would think that there is a great deal of resistance to deal with here." Dr. Vaupel concurred and added that "it would be useful to have a directive that would help accelerate progress toward more analysis."

Up to this point, Rep. Ritter had heard exactly what he wanted to hear. The next witness, however, Dr. Howard Raiffa of Harvard, head of the National Academy of Sciences Study Group on Risk and Decisionmaking, made explicit what the previous two speakers had demonstrated in spite of themselves, that risk analysis is not the panacea that its boosters claim it to be. He stressed the complexity of the decision-making process and listed some of the important factors that do not lend themselves to numerical resolution: How is one to calculate public discontent and personal happiness? Equity and justice? Individual freedom and personal autonomy? How is one to measure efficiency versus equity? Short-term versus long-run consideration? Man's interest versus nature's? The United States versus the world's? And so forth. He also raised a point that calls into question the value of the entire risk-analysis enterprise: "The probabilities of various adverse health or environmental effects are often so uncertain that experts disagree by orders of magnitude. . . ."

This fact was emphasized by a following witness, David D. Doniger of the Natural Resources Defense Council. He noted that studies conducted by OSHA on the carcinogenic effects of vinyl chloride have shown a range of disagreement of more than a millionfold. Cost estimates are also subject to large differences of opinion. The estimated cost of complying with a new standard for coke-oven emissions was ten times higher when prepared by industry experts than when calculated by government specialists.

Unfortunately, such uncertainties are usually most prevalent in matters of the greatest concern, such as PCBs, asbestos, radiation, and the chlorinated hydrocarbons in drinking water. This was emphasized by the next speaker, Dr. Nicholas A. Ashford of M.I.T.'s Center for Policy Alternatives. "The easy cases," he said, "which exemplify the benefits of simple risk assessment, aren't the ones that give us problems; those aren't the cases that need working on."

About halfway through the first day's hearings, the emphasis began to shift from matters of technical fact to issues of political and social philosophy. Speakers from the Monsanto Company and the National Association of Manufacturers commended Rep. Ritter and supported his initiative. In response, a representative from the United Steel Workers said that he was tired of being told "that we are safer in a steel mill than we are in our own bathtub," and characterized the proposed law as "a barrier against further expressions of responsibility."

Finally, Rep. Ritter, like a host whose party is being spoiled by unruly guests, expressed consternation: "I think we are seeing here somewhat of a polarization. I think people have come into this room with certain values locked onto certain interests. I would hope . . . that we could somehow get away from that."

Mr. Ritter's hope, however, was not to be realized. As the

afternoon drew to an end, the hearings increasingly deteriorated into bickering between industry spokesmen on the one hand and representatives of labor and public interest organizations on the other.

The following day's program, devoted mainly to spokesmen for various government agencies, was to see Rep. Ritter's discomfort grow apace. First there appeared a representative from the Office of Science and Technology Policy to say that his organization wanted no part of the responsibility that would be delegated to it by the new bill, and indeed did not support the concept of centralized control over the procedures of the regulatory agencies.

"It seems to me," commented Rep. Ritter, "that this country could be a lot better off just with a little bit of additional rational decision processes brought about by an increased involvement on the part of the scientific community. . . ." He spent the rest of the afternoon hearing his ideas rebutted by distinguished scientists.

Dr. Joseph V. Rodricks, a chemist with the Food and Drug Administration whose *curriculum vitae* runs to eleven pages, observed that the proposed legislation did not take into account the differences between the legislative acts under which Congress established the several regulatory agencies, and the different mandates that make a standardized procedure unwise, even within each agency. For example, the FDA regulates food additives, food contaminants, and drugs according to three completely different standards. Food additives must be absolutely safe, and the burden of proving the safety of an additive rests with its sponsor. For a food contaminent (such as PCBs and aflatoxins) the burden is on the government to prove hazard to health, and acceptable levels must often be established on the basis of distressingly meager information. In drugs, the balance of benefit and risk presents an entirely different sort of problem. Congressional statutes are full of such terms as "generally recognized as safe," "reasonable certainty," "unreasonable risk," "reasonable, technologically practicable and appropriate"—terms that differ subtly but substantially from one another, and do not lend themselves to a unified mathematical standard.

"No simple rules govern such complex problems," declared Dr. Bailus Walker, Jr., of OSHA. Dr. Peter W. Preuss, representing the Consumer Product Safety Commission, put the same message in more dramatic terms: "When comparisons are made among different types of risks, the uncertainties become awesome. . . ." Although statistical methods clearly are useful, "the commission's overall approach for setting priorities is to rely on its collegial judgment rather than on any set formula." *Collegial judgment:* I liked the phrase.

In the Food Safety and Quality Service, according to spokesman Dr. Thomas P. Grumbly, risk assessment is used as an aid to decision-making, but its limits are recognized. "In many ways," according to Grumbly, "it is no more precise a tool than its nineteenth-century precursor, Jeremy Bentham's pleasure-pain calculus."

The final witness, Dr. Richard Dowd of the Environmental Protection Agency, spoke in the same vein. "The diverse nature of the EPA's responsibilities," he observed, "would argue against changing the laws to set a common framework for regulation."

If Rep. Ritter found the opposition to his bill at all convincing, he gave no indication of this in his final remarks. "I would just like to say," he concluded, "that Congress and the American people that we represent in Congress are beginning to rebel against what they perceive to be an overly bureaucratic and dogmatic and perhaps dictatorial regulatory apparatus."

Although the hearing ended with expressions of thanks and mutual admiration, it was clear that the proposed legislation had provided a battleground for forces that were bitterly antagonistic.

* * *

I approached the subject of risk assessment with high hopes and a disposition to be persuaded. But after reading the transcript of the subcommittee hearings I could only conclude that Rep. Ritter's attempt to apply mathematical consistency to the regulatory process was deplorably simplistic. Worse than that, it provided an opportunity for the industrial enemies of regulation to take "rationality" as their rallying cry, thus giving force to the long-standing charge that engineers are flunkies for the corporate establishment.

Clearly, the public perception of risk, and the actual management of risk, are matters that deserve our keenest attention. Responsible experts in the field, however, consistently warn that risk assessment is a delicate tool that needs to be applied sparingly, not a machete to be flailed against the supposedly overgrown regulatory jungle.

Of course, regulators need to be regulated. Each regulatory agency is properly constrained by the language of the statute under which it operates, by continuing legislative oversight and judicial review, and by the political realities of the day as perceived by the canny bureaucrats within the agencies. Add to this the judicious (though not compulsory) use of risk analysis and we have the basis for that "collegial judgment" upon which the regulatory process should properly be founded.

On June 17, 1981, the Supreme Court ruled that in promulgating a cotton-dust safety standard, OSHA was not required to demonstrate a reasonable relationship between costs and benefits, since Congress had mandated a standard that would protect health "to the extent feasible." "For deregulation, a detour," grumbled *Business Week,* but "not a roadblock." In a dissenting opinion, Justice Rehnquist complained that by using such vague language Congress simply abdicated its responsibility for the making of a fundamental and most difficult policy choice. "I have no doubt," he noted, "that if Congress had been required to choose whether to mandate, permit, or prohibit the secretary from engaging in a cost-benefit analysis, there would have been no bill for the president to sign."[3] A telling point, yet in passing the ultimate decision along to the bureaucracy, Congress was in fact making a considered policy choice.

On September 24, 1981, at a brief hearing before the Subcommittee on Science, Research and Technology of the House Committee on Science and Technology, the Ritter legislation suffered a serious rebuff. Dr. James C. Miller III, Chairman of the Federal Trade Commission and former Director of the Presidential Task Force on Regulatory Relief, announced that, in view of other anti-regulatory actions already being taken, the Reagan Administration considered the Ritter bill to be "basically redundant." This signaled defeat for the bill, though not for the ideas behind it.

Just as the beginning of wisdom is self-knowledge, so is the essence of engineering rationality a recognition of its proper limits.

17

Risk and the Loss of *Challenger*

No discussion of engineering and risk would be complete without reference to one of the most shocking technological mishaps of our time, the explosion of the space shuttle *Challenger* on January 28, 1986. From an engineer's point of view, the loss of *Challenger* raises compelling questions which will continue to haunt the profession long after the nation's bewilderment and grief have faded.

In some respects the disaster was merely a classic example of engineering failure, an object lesson worthy of study by future generations of designers. Its emotional impact was extremely high, but other tragedies—failed dams, collapsed buildings, burned nightclubs—have also had their quotient of horror and sadness. What makes the event unique is the drama of the decision-making process and the ensuing debate about it in a national forum.

First the classic elements: Once again a minute detail undid a grandly conceived technological project. The failure was traced to a seal between two segments of a booster rocket, or more precisely to a rubber gasket that was an integral part of that seal. Once again it was demonstrated that engineers must deal with real materials—real rubber, putty, and steel—not ideal materials whose behavior can be confidently predicted by scientific formula. The performance of these real materials in manufactured products can only be ascertained by testing, and sometimes it seems as if no amount of testing can ever be "enough."

We discovered anew that in complex engineering systems the possibilities of failure are compounded by a multiplicity of potential coincidences. Just when *Challenger*'s primary O-ring gasket failed, allowing hot gasses to escape, a second adjacent O-ring, designed originally for redundant safety, was unseated from its groove by the movement of the rocket casing under pressure. Small wonder that engineers speak of Murphy's Law. As Herbert Hoover wrote in his memoirs, failure for the engineer "is the phantasmagoria that haunts his nights and dogs his days."[1]

Of course, failure has always been an integral element of technological advance. When the first jet airliner, the British Comet, was put into service in 1952, it had been test-flown for almost three years. All went well until early 1954 when a plane inexplicably crashed into the Mediterranean near Naples. After being grounded for three months, during which some fifty modifications were introduced to anticipate every conceivable danger, the Comet fleet went back into service, only to have a second plane disintegrate in midair. Eventually the fault was found to lie in fatigue cracks in the corners of windows, a problem that at the time had seemed unthinkable in such relatively new machines. The space shuttle is more than just a new product; it is practically a new technology, and viewed in the light of history, some sort of accident was almost to be expected.

But the loss of *Challenger* differed from the Comet crashes, and other similar calamities, in that a number of engineers had predicted that it might happen and strenuously objected to the launch. The engineers at Morton Thiokol, Inc., manufacturer of the rocket boosters, knew that cooling reduced the resilience of the all-important O-ring gaskets, making them less efficient in sealing the gaps they were required to seal. On an earlier flight, in which the seal temperature had been fifty-three degrees, there was evidence that some gas had escaped past one of the rings. On the evening of the tragic flight, as temperatures at the *Challenger* launch site dropped into the low twenties, fourteen engineers at the Thiokol plant in Utah unanimously concluded that the flight should be postponed: the possible effect of the cold on the O-rings, they agreed, created an unacceptable level of risk. This recommendation was transmitted to NASA officials and a telephone conference was set up to discuss the problem. The conference quickly became a confrontation. George Hardy, NASA's Deputy Director of Science and Engineering, stated that he was "appalled" at the reasoning behind Thiokol's stance. Lawrence Mulloy, NASA's Chief of Solid Rocket Booster Program, protested that Thiokol was changing the flight criteria on the night before a scheduled mission and exclaimed, "You can't do that."

After a lengthy debate, a five-minute recess was called, which stretched on for half an hour. At Thiokol, while the fourteen engineers remained unanimous in their opposition, Senior Vice President Jerald Mason declared, "We have to make a management decision," then asked Vice President for Engineering Robert Lund to "take off his engineering hat and put on his management hat." Subsequently, Mason and Lund, along with Joseph Kilminster, Vice President for Booster Programs, and Calvin Wiggins, Vice President for Space Projects, approved the launch and so notified NASA.

Much of the commentary after the event stressed that the findings of "engineers" had been overruled by the judgment of "managers." According to *The New York Times,* "Company managers testified that the engineers had played no part in that decision, which was based on a poll of four Thiokol managers." When Mason was asked what he had in mind when he told Lund to replace his engineering hat with a management hat, he said that he meant "it was going to take a judgment rather than a precise engineering calculation."

At first hearing, these comments might give comfort to anyone concerned with the good reputation of engineers, but I am not comforted. The implication that engineering consists of precise calculation while "judgment" is left to another class of person is inaccurate; further, it is demeaning to a profession that has always stressed art, imagination, and wisdom at least as much as exactitude. There are, as I have argued, many instances in which engineering facts can and should be isolated from business or political choices, but the *Challenger* launch was not one of these.

The four so-called Thiokol "managers" are, in fact, engineers. Mason has a degree in aeronautical engineering, Lund in mechanical; Wiggins has a degree in chemistry, and Kilminster a master's in mechanical engineering on top of an undergraduate degree in mathematics. The two NASA "officials," Hardy and Mulloy, who urged that Thiokol approve the launch, are also engineers, as are the key NASA people above them. These men were educated as engineers and had worked as engineers, eventually moving into positions of executive responsibility. They did not thereupon cease being engineers, any more than a doctor who becomes director of a hospital stops being a doctor.

Were these engineer-executives under pressure to meet a launching schedule? Of course. But pressure is inherent in engineering. It has often been said that engineering is doing for one dollar what any darn fool can do for two. It might with equal truth be said that engineering is doing in one year what any darn fool can do in two. Considerations of economy and schedule are inherent in all complex engineering projects. Given unlimited funds and time, the shuttle design could surely have been improved in many respects, but the engineers at NASA and Thiokol had no reason to expect either a blank check or freedom from a sense of urgency. Pressure goes with the job like the proverbial heat in the kitchen. It may help explain, but it cannot excuse, an engineering mistake.

On the other hand, after the event it was not reasonable for the public to say "safety first" while ignoring the other imperatives driving the space program—national defense, scientific exploration, technological progress, and even such intangibles as community morale, all to be enhanced in a hurry for an affordable price. If safety was the overwhelmingly dominant consideration, we would never blast human beings into space on the top of rockets in the first place. (There are some people who argue that all space projects should be unmanned, but they are clearly in the minority.) Richard Truly, the new shuttle chief appointed after the disaster, promised a more "conservative" flight philosophy, but he put the dangers of the shuttle program in perspective: "The business of flying in space is bold business. We cannot print enough money to make it totally risk-free."[2]

All we can fairly say is that safety was and is one very important aspect of the space program—safety not only for the astronauts but for the public at large. (The errant booster rocket was destroyed by remote control for fear that it might fall in a populated area.) Numerous procedures were established long ago by NASA to assure safety, one of the most important being the required "sign-off" before launch by each major shuttle subcontractor. The system is sound and it should have worked. Fourteen engineers at Thiokol recommended a postponement. Incredibly, four of the top executive-engineers at Thiokol, pressured by administrator-engineers at NASA, ignored the warnings and agreed to proceed. This was bad engineering as well as bad management.

There were no precise facts to rely on. The rubber O-ring gaskets had not been tested under flight conditions at temperatures as low as those prevailing on the fateful night. But the engineers knew that the colder it got, the less pliable the rubber and the more likely the O-rings would be incapable of containing the hot gasses. One or two apprehensive engineers might properly have been disregarded. As a newspaper editorial observed after the tragedy, "with hundreds of engineers reporting, NASA cannot allow the most anxious among them to veto a flight."[3] But the unanimous recommendation of the Thiokol engineering staff should have carried the day.

The presidential investigative commission eventually recommended that an independent safety panel be established within NASA to make sure that safety considerations are adequately represented in future decision-making. This is not a bad idea.

Safety engineering has become something of a specialty, and many large corporations isolate the safety function in the organizational structure. But any such change, whatever its merits, should not obscure the fact that the pre-existing system was not seriously flawed. The annual report of the Aerospace Safety Advisory Panel, written shortly before the *Challenger* accident, expressed general satisfaction with NASA's attention to key safety concerns. The calamity stemmed not from an inadequate system but from human error.

Should the junior engineers at Thiokol have "blown the whistle," gone out of the chain of command, or otherwise publicly aired their concerns? I cannot fault them for doing no more than argue vigorously. They had no clear evidence, only clues and uneasy feelings. Most important, they were not reporting to politicians or non-technical administrators, but rather to sophisticated engineers like themselves, four in their own company and, to their sure knowledge, at least three at NASA.

The ironic truth is that if the Thiokol managers and NASA administrators had not themselves been technical experts they would never have dared approve the launch. Don K. Price, former dean of Harvard's School of Public Administration, has observed that "on the undefinable boundary between a profession and administration, the professional has a tactical advantage." For administrators or politicians to go against the consensus of either a scientific discipline or a profession "is extremely hazardous."[4] In his book *The Scientific Estate,* Price speaks of a social spectrum along which are situated four "estates": scientists, professionals (including engineers), administrators, and politicians, each with its own bases of power and ability to influence the others. Politicians usually have ample room to maneuver because technical experts are so often divided in their opinions. But in the case of a shuttle launch, no sane politician or administrator would go ahead without solid professional support. Public opinion would not tolerate it. This is why the White House was at such pains to deny that it in any way attempted to influence NASA's decision. And this is why it was such a serious error for the NASA engineer-executives at Marshall Space Flight Center not to tell their own superiors about the concerns of the Thiokol engineers.

Whatever future institutional changes may be conceived at NASA, there will always be risk decisions to be made in the midst of uncertainty. In the post-accident hearing, there was talk not only about separate safety panels, but also about new techniques of risk assessment. It is tempting—particularly after a dramatic failure—to pin one's hopes to a new methodology.

But no technique or table or organization can take the place of the collegial judgment that, as noted in the previous chapter, was advocated during the hearings on Congressman Ritter's bill. Such judgment was manifested in the unanimity of the fourteen Thiokol engineers the night before the ill-fated launch. Perhaps the *Challenger* tragedy will help restore respect for this time-honored aspect of professional practice.

18

The Civilized Engineer: (1) The Concept

Herbert Hoover said that during his lifetime he had seen engineering change from a trade into a great profession. This was not completely accurate since the transformation of the typical engineer from independent consultant to corporate employee, which occurred during his years, can hardly be viewed as professional growth. Nevertheless, he had seen the foundations of engineering change from rule-of-thumb experience to formal science, and the apprenticeship system replaced by accredited university curricula—including, wrote Hoover in his memoirs, "wider learning in the humanities."[1] He also saw impressive growth in the professional societies and, along with other engineering leaders, he helped to elevate professional standards. Established in his early years as a successful mining engineer, Hoover gained renown leading American relief operations in Europe during and after World War I. This led to his appointment as U. S. Secretary of Commerce and eventually to his election in 1928 as president of the United States. He authored many technical papers, an important and successful textbook, and a number of well-written articles explaining engineering matters to the public. In collaboration with his wife, he translated Agricola's *De Re Metallica* from the Latin.

What would this scholar/public servant/engineer think of his profession today? Would he perhaps fear that it was reverting from a profession back into a trade? He would be impressed, surely, by the growth of engineering science and by the wonders of contemporary technology. But how would he feel about the increasingly vocational nature of engineering education and about the constricted technical character of much engineering practice? Would he not wish to regenerate a sense of pride, intellect, and cultural sensibility among his co-professionals? I believe that he would. And in seeking to bring engineering to the highest possible level he would find himself in the company of many other engineers and engineering educators who are proud but concerned, self-assured but dissatisfied, wanting more for their profession than the situation that now prevails.

Whatever its problems, past or present, the engineering profession in America has not suffered from want of attention. Although it sometimes appears as if we lurch from crisis to crisis —and indeed our political behavior often reflects the human tendency to concentrate on the problems of today rather than worry about tomorrow—the foundations of technological competence have received much earnest study and support. From the nation's early days, effective government enterprise has supplemented far-sighted private initiatives.

The U. S. Military Academy was created in 1802 mainly to provide a professional education for home-grown engineers. The founding of the American Society of Civil Engineers in 1852 marked the beginning of formal organization among American technologists and the flowering of professional self-awareness. The Morrill Act, passed in 1862, established the land-grant colleges, the main purpose of which was to provide instruction in the mechanical arts and agriculture. The following year Congress issued a charter to the National Academy of Sciences, a private, non-profit, self-governing membership corporation. (The National Academy of Engineering and the Institute of Medicine became distinct affiliated entities in 1964 and 1970, respectively.) Although the Academy has retained its status as a private organization, it performs services for the government, primarily through studies commissioned by various federal agencies. The National Research Council, created in 1916, is the operating arm through which such studies are performed.

The National Science Foundation, founded in 1951, is an agency of the federal government whose primary function is to support worthy scientific and technical enterprise, primarily through academic grants. The Office of Science and Technology Policy is part of the executive branch of the government; the Office of Technology Assessment serves the Congress. These and a multitude of other government agencies, academies, professional societies, foundations, and industrial associations make it their business to monitor, shape, and support engineering in America.

The most comprehensive recent examination of the profession was conducted from 1983 to 1986 by the National Research Council through a Committee on the Education and Utilization of the Engineer. Entitled "Engineering Education and Practice in the United States,"[2] the study was funded by the National Science Foundation, the Departments of the Army, Navy, and Air Force, NASA, and the Department of Energy. Additional grants were provided by Eastman Kodak, Exxon, General Electric, IBM, Lockheed, Monsanto, and the Sloan Foundation. One element of this multifaceted enterprise was a three-day seminar held in Washington, D.C., featuring papers presented by numerous historians of technology. As a member of one of the subcommittees that participated in the project, I attended the seminar and, thinking back on the occasion, recall basking in an almost uninterrupted testimonial to the achievements of my profession. I don't believe the historians set out to praise engineers, but the facts spoke for themselves. Whenever the nation has called upon its engineers, they have responded magnificently. For example, when the supply of German synthetic organic chemicals was cut off during World War I, research laboratories and manufacturing plants were speedily established, and academic programs in chemical engineering proliferated. While additional chemical engineers were being educated, scientists and engineers from other disciplines stepped into the breach. Not only were the needs of the nation attended to, but in short order the United States became a worldwide leader in the field.

The early days of the petroleum industry reveal similar skill, adaptability, and cross-germination between specialties. In 1920 there were only two university programs in America for the training of petroleum engineers. However, mining engineers and geologists, along with other engineers and scientists, did the work that needed doing. The supply of petroleum engineers gradually increased to meet demand. The 1973 oil embargo created another crisis that was met in the short run by engineers who moved into the field from related specialties.

Other successful responses are to be found in the development of weapons, particularly in the atom bomb Manhattan Project, and in the fabulous Apollo lunar-landing program. Transport, water supply, sanitation, communications, manufacturing, building construction, computers—it is difficult to think of an expressed societal need to which engineers have not responded with ingenuity, resilience, energy, and dedication. If engineers have not created Utopia, or even mapped out the road to that fabled land, this has less to do with their inadequacies than with the ever-changing objectives of our society as a whole. The competence of American engineers simply cannot be questioned. It has been a significant national resource—indeed, a national treasure—arising from wellsprings of native ingenuity and nourished by concern at the highest levels.

In its development, the American engineering profession has drawn upon two competing yet complementary traditions: the hands-on, muddy-boots pragmatism inherited from Britain and the elite, science-oriented approach of the French *polytechnique.* Science and mathematics gradually have taken a central position, with emphasis being placed upon their creative application. The less theoretical "hardware" aspects of technology have been delegated in part to graduates of two-year technician courses and four-year programs in engineering technology. Pure science has been left to the scientists. Business management has developed into a discipline of its own, although economic parameters remain central to an engineer's tasks.

But while engineers have performed miracles, and seen their profession grow to ever higher levels of knowledge and skill, it has not been without cost. The increasing technical content of even the most elemental engineering education has inevitably changed the nature of that education, forcing out peripheral concerns and all but eliminating the liberal arts that once were the heart of the college experience. Inescapably, engineers have become less broadly educated, less wide-ranging in their background and interests—in a word, less "cultured."

Engineering education in America came into being as undergraduate education, and its basic degree has always been the four-year bachelor of science. This differs from other fields—notably medicine, law, and business administration—in which a general bachelor's degree is followed by two or more years of specialized graduate school. Many authorities have recommended that engineering adopt a minimum five-year course for entry into the profession. This has been tried from time to time at several institutions, but except for the Thayer School at Dartmouth College, all have reverted—under the pressure of economic competition—to the four-year program. A few engineering students voluntarily earn a bachelor of arts degree prior to the engineering B.S. But the vast majority take no more than a half-dozen non-technical courses during their undergraduate years, and of these most are likely to be in the "useful" social sciences rather than the "useless" humanities such as literature, history, and philosophy. (Post-graduate programs, pursued by about a quarter of engineering graduates, are, of course, totally technical.)

A century or more ago it was possible to combine liberal learning with technical training in a four-year engineering curriculum. There were few texts, a shortage of qualified professors, and no great abundance of engineering substance to be taught. Moreover, early engineering educators—Benjamin Franklin Greene at R.P.I., William Barton Rogers at M.I.T., Andrew Dickson White at Cornell—perceived the liberal arts as an integral part of any engineer's education. In 1867, one-third of M.I.T.'s mechanical engineering program consisted of languages and humanities. However, with the exponential growth of engineering knowledge and the expanding technological needs of a young nation, the situation changed rapidly. Technical content increased, proponents of the four-year program stood firm against expansion of the course, and liberal education was perforce squeezed out. Eventually, to forestall complete vocationalizing of engineering education, leaders of the profession agreed upon a minimum 12½ percent liberal arts component for accredited curricula. This translates into the six non-technical courses I allude to above.

For a while even this meager portion provided the young engineer with a slight introduction to the general culture, especially since broad survey courses such as Western Civilization or Great Works of Literature were widespread. Doubtless this is what Herbert Hoover had in mind when he wrote about "wider learning in the humanities." But these courses fell into disrepute, scorned by students during the rebellions of the sixties, and readily abandoned by academicians who regarded them as superficial and lacking in rigor. Concurrently, the social sciences took over an ever larger portion of the liberal arts curriculum. So the plain truth is that most American engineers do not read the "great books," do not study the history of their nation or any other, and are effectively cut off from humanistic, philosophical, civilizing education.

While the scientific content of engineering grew, and the liberal element of engineering education diminished, the profession increasingly attracted the sons of rural and blue-collar families, young men to whom the arts and humanities held little appeal. Psychological research indicated that engineers were becoming "narrow" in personality and outlook. In the 1950s an analysis of several studies of engineers revealed an unsettling profile:

> Constricted interests are apparent in their relative indifference to human relations, to psychology and the social sciences, to public affairs and social amelioration, to the fine arts and cultural subjects and even to those aspects of physical science which do not immediately relate to engineering.[3]

At the same time, the vast majority of engineers entered the ranks of industry as employees of large corporations and the percentage of engineers in private practice dropped accordingly. The "status" of the profession declined.

I regret the passing of the venerable engineers of the nineteenth century, the Roeblings and their ilk, men of learning and artistic sensibility, scholars as well as doers, leaders who gave the profession an aura of true nobility. Herbert Hoover and a few others carried this tradition into the middle years of our own century, though such patricians have become increasingly rare. I dream of their return, and I intend to plead the cause of a humanistic professionalism, an ennobled engineering that will rise out of the ashes of vocational training. Before I do this, however, let me give due credit—indeed, let me pay homage—to the "ordinary" American engineer of the past one hundred years.

One need not be a broadly educated scholar in order to be a topnotch engineer. The evidence is clear: The technical feats recounted at the seminar I referred to a moment ago were accomplished mostly by engineers of little liberal education and, in some respects, in defiance of ancient cultural values. The humanities as we know them hailed from Europe, and a century ago Americans had good reason to question their worth. The Old World had created a civilization rich in beauty and intellect, but only within a framework of rigid class structure, religious intolerance, and perpetual war. Traditional liberal education did not so much "liberate" men and women as it defined those who were rich, held political power, and dealt unjustly with their fellows. As the American frontier was pushed westward, and as hordes of immigrants arrived in the New World fleeing poverty and oppression, there was little place for the niceties of Continental culture. The schoolhouse and the Bible traveled with the pioneers, but respect for learning was infused with respect for hard work and technical ingenuity. Where classical learning did take hold—mainly in the Eastern universities—the old Platonic scorn for craftsmanship, imported from Oxford and Cambridge, served to alienate engineers and inspire them to seek alternative values. As the engineering sciences became in themselves richer and more intellectually demanding, engineers developed new self-esteem and began to challenge time-honored definitions of "higher education." Is not an engineering problem as complex, intellectual, creative, and existentially fulfilling as a sonnet or a philosophical theorem? And is not culture, isolated from useful accomplishment, a false measure of social superiority?

As a network of railroads spread across the continent, soon to be followed by mines, factories, cities, and an electric power grid, engineers took on the aura of heroes. An indication of how they were viewed can be garnered from a passage that appeared in a popular novel of 1911:

> It was the last night out. Supper was over and the men, with their pipes and cigarettes, settled themselves in various careless attitudes of repose after the long day. . . . All were strong, clean-cut, vigorous specimens of intelligent, healthy manhood, for in all the professions, not excepting the Army and Navy, there can be found no finer body of men than our civil engineers.[4]

It is not surprising that engineers proudly came to think of themselves as men of action. (I say men advisedly, because practically none of them were women.) This robust image, developed while building railroads and drilling for oil, followed them into factories and mills and even into laboratories and drafting rooms. They fancied themselves doers, not readers of poems or admirers of paintings. In 1916 a professor of English at M.I.T. observed that engineers viewed "culture" as "a pose, impressive perhaps with some women, but scarcely effective among men."[5] As late as 1956 a professor at City College in New York noted that engineers tended to be "rough, tough spirits" who took "pride in cultivating construction camp and barroom manners rather than the deportment that would grace a drawing room."[6]

If this was the feeling in Boston and New York, one can imagine what it was like farther west, where many of the nation's engineers were raised and got their schooling. (Into the 1960s rural areas and small towns produced more than half of America's engineers; when large cities contained one-third of the population they were producing only one-sixth of the engineers.) Asked why it was that small-town boys from the Midwest dominated the engineering frontiers, Robert Noyce, co-inventor of the integrated circuit, who grew up in Grinnell, Iowa, said it was because of necessity. "In a small town," he explained, "when something breaks down, you don't wait around for a new part, because it's not coming. You make it yourself."[7]

High culture meant little to small-town boys. Yet many of them became great engineers. Tom Wolfe has observed that most of the major founders of Silicon Valley were reared and went to college in small towns in the Midwest and West. As for the Apollo Project, notes Wolfe, "It was engineers from the supposedly backward and narrow-minded boondocks who provided not only the genius but also the passion and the daring that won the space race. . . ."[8]

In the big cities, too, boy tinkerers turned engineers found little appeal in the liberal arts. I have reported how in college I

—a city boy—was swept up in the fascination of mathematics and the engineering sciences. But for all the narrowness of their interests and the limits of their education—and probably in some cases *because* of their intense focus—American engineers have done marvelous deeds. Even the psychological inadequacies revealed in numerous studies—and reaching new extremes in the computer "hackers" of today—have played a positive role in sublimated creativity. Surely engineering work has provided satisfactions for people whose lives might otherwise have been filled with frustration. In turn, obsessed technologists have repaid society many times over for the opportunity to follow their own particular paths of interest.

Most engineers have found a degree of fulfillment in their profession that is, or should be, a source of envy for others. A recent poll shows that 75 percent of electrical engineers are "quite satisfied" or better with their working lives. A second poll indicates that only 9 percent are less than "moderately content." More impressive still, the polls reveal that such discontent as there is stems mostly from employment conditions that prevent individuals from doing real engineering work.[9]

As for honesty, conscience, and good citizenship, I believe that engineers stand up extremely well when compared with other groups, particularly groups such as lawyers and business executives, who have had the benefit of a traditional liberal arts education. There does not appear to be any correlation between morality and book learning. I have met hundreds of engineers and engineering students who for all their unfamiliarity with literature, history, and the arts, are good, productive, responsible members of the community. Small wonder that so many engineers and engineering educators resent being derided for lack of "culture" and resist attempts to increase either the length of the engineering college course or its liberal arts component. The system works as it is, they maintain, and who can gainsay them? Society is well served and can ill afford to spend more time or resources on getting the engineers it so desperately needs. Industry is well satisfied. And prospective engineers, anxious to get on with their careers, are content. Besides, comes the impatient retort, in a technological age it is the technologically uninformed who are the true know-nothings.

Nevertheless, there are widespread feelings that something is amiss. Even if the system does "work," and even if most engineers and employers of engineers see little reason to change, a large and vocal minority are not at all satisfied.

Uneasiness about the vocational emphasis in engineering is nothing new. In the 1860s an R.P.I. professor expressed apprehension about a "professional learning" that would be a "mere excrescence on an otherwise ill-developed intelligence, rather than a crowning dome upon the capacious body of a well-proportioned mental structure."[10] In 1902 the president of Stanford complained, "We can not make an engineer in four years if we do anything else with him, and there are very many things besides engineering which go to the making of a real engineer."[11] By 1920, Charles Steinmetz, the great electrical engineer and star of GE's laboratories, was calling for the "broadness of view and common sense which only a broad and general education can give." He warned that "special training rather tends to narrow the view and incapacitate the man from taking his proper position as a useful member of society."[12]

In recent years, two things have happened to give such misgivings added urgency. The first is practical and relates to the need for engineers to be able to "communicate" in order to be effective. Paradoxically, as the technical content of engineering has grown, so has the need for engineers to verbalize—to speak and write clearly—and work with other people on a non-numerical plane—to listen, understand, explain, persuade, and empathize. Engineering is more a group enterprise—a team effort, if you will—than it used to be. According to the president of the Bethlehem Steel Company, engineers "must be able to logically present their information, thoughts, and discussions, both orally and in writing to others. The best idea in the world is worthless unless it can be 'sold' to others."[13] A vice president of the Dow Chemical Company endorses this concept and notes that it applies to situations beyond the corporate walls: "The ability to write and speak clearly with people outside of your particular discipline is becoming more and more a requirement for success or survival. The technical graduate entering industry today will, in all probability, spend a portion of his or her career explaining technology to lawyers—some friendly and some not—to consumers, to legislators or judges, to bureaucrats, to environmentalists and to representatives of the press."[14]

Most employers of engineers agree with the executive vice president of Deere & Company who says that "the engineers coming into our industry today are the best we've ever had," but goes on to note that in the future, "we'll need engineers who are articulate, who can write and speak concisely, effectively and persuasively [and who will] not be subject to cultural shock if involved in projects in the Middle East, the countries of the East Bloc, or even the People's Republic of China."[15] Again and again one hears the message repeated: More is and will be required of engineers than technical skills. A recent study by AT&T found that engineers are only half as likely to attain senior management status as liberal arts graduates.[16] The Edison Electric Institute surveyed twenty public utilities and found a nearly unanimous call for engineers with a broader background in the humanities.[17]

A certain amount of this clamoring for well-rounded engineers is mere lip service. I once heard a highly placed executive say in a moment of candor, "I want one engineer out of twenty to be eloquent and imaginative. I can't use a whole division full of hotshots!" As an employer of engineers myself, I knew exactly what he meant. But one in twenty or nineteen in twenty—whatever the appropriate ratio—there is clearly a need for more engineers with richer experience in writing, reading, speaking, and thinking in broad, humanistic terms.

The other source of heightened interest in liberal arts for engineers is the fear and suspicion of technology which emerged with the development of nuclear weapons, and intensified in the wake of the environmental crisis of the 1970s. Although the neo-conservative mood of the 1980s has seen an abatement of anti-technological sentiment, the feeling has grown that engineers should be aware of the social consequences of their work. Sensitized and enlightened engineers, it is claimed, will create technological products that are environmentally benign, aesthetically pleasing, and socially beneficial. I put the argument this way almost twenty years ago:

> Someone must step forward to say, "We can afford to make that automobile a little safer," "Let us build a factory that is more attractive," "Let us consider the possible harmful effects of that insecticide before we market it," "Let us develop a plant process that will not pollute the water we use," "Let us make that machine a little quieter," "Let us not demolish that historically precious old building," "Let us locate that dam, not only where it will generate the most power but also where it will serve the interests of the community—aesthetically, politically and socially. . . .
>
> This "someone" cannot be an ordinary citizen of good will. He must be able to bolster his arguments with facts—technical, scientific, and economic. Hunches and sentiments will not be sufficient. His recommendations, in order to be persuasive, must be founded in a knowledge of resources, materials, and energy conversion; statistics, probabilities, and decision theory; computers, controls, and systems engineering. Moreover, this "someone" must be concerned. He must be articulate. He must be esteemed. And he must have a highly developed moral

> and aesthetic sensibility. In short, he must be a liberally educated engineer.[18]

As I have made clear in subsequent writings—and in earlier chapters of this book—I now believe such a view to be simplistic. Partly this is because the world has changed. Technology is much more democratically controlled than it was when I wrote those words. The control comes from new laws, new regulations and regulatory agencies, new legal concepts concerning liability, new legal constraints on the self-government of professionals, new interest and study by the press and media, and vastly increased concern and involvement by the general public. Partly, also, I have learned by living and by reading a bit of history that the tastes, prejudices, and moral beliefs of individual engineers are not the decisive factors in the shaping of technology. Finally, I have found that the linkage of a liberal arts education to social conscience and political awareness is not as direct as one might think. As President Derek Bok of Harvard has admitted: "Formal education will rarely improve the character of a scoundrel."[19] Obviously people with no college education at all can have superior ideas about how life should be lived and more particularly about how technology should be used.

Having said all this, I am nevertheless convinced that the quality of our technology, and consequently the quality of our lives, would be improved by the liberal enrichment of engineering education, by the broadening of horizons, the deepening of cultural awareness—in short, by the civilizing—of engineers.

Engineers are already civilized, to be sure. More than this, engineers are civiliz*ers,* since civilization in its most elemental sense means bringing humans physically out of the state of primitive savagery. Engineers are also among the most civil of citizens. But if they do not partake of the most sublime creations of civilization—the arts and intellectual discourse—they are not as civilized as they might be. And if they have talent and intelligence (as engineers do have in full measure) then they are not so much uncivilized as they are anti-civilization. Of course, the same can be said of "intellectuals" who ignore the great achievements of science and technology.

In trying to comprehend this elusive topic, the metaphor of organic growth seems to me particularly apt. The "roots" of a civilized society are the technical accomplishments that relieve people of brute effort and make humanity possible. When we speak of the "fruits" of our efforts, of the "flowering" of civilization, we refer to art, philosophy, and science. If the fruits and the blossoms are not returned to nourish the soil, then life loses strength and its flowering becomes less radiant. Which is to say that if engineers exclude themselves from the grand cycle, if technology is not enriched by new beauty and insight, then the growth that follows is less luxuriant and all of humanity is the loser.

The resort to metaphor implies the inability to adduce proof, and I confess that in part my argument is intuitive. Preceding all specific rational arguments is the emotional conviction that the more liberally educated our engineers the better will be our society as a whole.

Several years ago there was quite a bit of interest in something called transcendental meditation, and I was deeply impressed by a statement made by its supporters. They claimed that if just one percent of the people in a community were meditating regularly, this would yield measurable benefits such as a reduction of mental illness and crime. The serenity of the meditators (so goes the theory) is diffused by personal contact. Cheer begets cheer, calm begets calm, politeness begets politeness, and so forth. Although I hold no brief for meditation as a source of serenity, I find the concept of beneficial diffusion both appealing and persuasive. By analogy, I see the liberal education of engineers as a source of amelioration for our technology and for our society. Enlightened, sensitive, broadly educated engineers—civilized engineers—would endow their work with an additional touch of grace, and in their contact with fellow citizens would create a more affirmative and profound feeling about engineering than that which exists today. The exact process by which this would occur might be obscure, but the ultimate benefit, I believe, is certain.

Descending from the level of long-term vision to that of short-term survival, I suggest that liberal education for engineers would bring members of the profession into leadership roles from which they are presently excluded, and that this, in turn, would bring significant economic and political benefits. More than a generation ago James Reston of *The New York Times* expressed the problem in these words:

> The politicians, and even the statesmen, are merely scrambling to deal with the revolutions in weapons, agriculture and industry created by the scientists and the engineers. The latter have transformed man's capacity to give life, to sustain and prolong life, and to take life; and the politicians no longer find that they can deal with all the new complexities and ambiguities. . . .[20]

Politicians are not the only ones who are perplexed and overwhelmed by the pace of technological change. The crisis extends to every area of society—industry, finance, education, the courts, and more. Engineers do not have answers to all the great questions, but their absence from decision-making groups clearly works to our nation's disadvantage. Their use as sometime "advisers" only partly compensates for their absence as principals. In Japan half the senior civil servants and half the industrial directors have engineering qualifications; in the United States the figure does not exceed 15 percent, and the engineers who make up this relatively small group are concentrated in technical positions.

The leaders of our society are almost exclusively men and

women who have received a liberal arts college education. This may be partly custom and snobbery. But it could hardly be otherwise. For it is liberal education that transmits tradition, history, ideas, broad understanding of the world—the shared knowledge and values that bind us together as a society. In this technological age our society needs technically knowledgable people in its highest councils, and engineers—or some of them, at least—should be trained accordingly. This impacts directly, I believe, on the ultimate well-being of the American commonwealth.

Also important for the well-being of society is the role that engineers can play in vitalizing and transmitting our cultural heritage. Large numbers of the brightest young people today want to be engineers. This is fine, but if becoming an engineer means turning one's back on literature, history, and the arts, then who will be left to nurture our great cultural inheritance? I do not want to leave culture to academic specialists and latter-day bohemians any more than I want to leave the running of the world to lawyers and MBAs.

As for engineers who are concerned about the social status of their profession, liberalized education certainly is a cause worth championing. It is galling to hear a public relations man refer to the typical engineer as "this kind of faceless, anonymous character,"[21] and to learn that three out of four journalists agree that engineers are "wooden."[22] To the extent that representatives of the profession become eloquent, informed, and broadly cultured, the problem of "image" will take care of itself. It is not just a matter of superficial refinement, equipping oneself for "parlor conversation." It comes down to recognizing that words and ideas are worthy of respect, and that important things happen in the parlors of this world as well as in its laboratories and offices.

And while we are thinking about the advancement of society and the reputation of the engineering profession, let us not forget the welfare of the individual engineering student. For the young engineer, liberal education will bring delight in the arts and an introduction to those insights of literature, history, and philosophy that so many engineers in their mature years regret having neglected. In an age when an increasing number of engineers will spend their working lives in large organizations, they need more than ever the personal discoveries and philosophical self-reliance that lie at the heart of the humanities. I am convinced also, as I tried to show in Chapter 2, that a humanistic education helps engineers to appreciate the satisfactions inherent in their *own* professional work.

Finally, I believe that liberal education, by expanding the intellect and exercising the imagination, can make engineers technically more able. The importance of written and oral communication in modern engineering work has already been noted. But beyond this, in the non-verbal creative recesses of the engineer's mind, I believe that flexibility and variety—poetry, art, music, stories, myths—have a constructive role to play. I recognize that this is difficult, if not impossible, to prove, and that a case can be made for the creativity of the obsessively single-minded engineer. Nevertheless, there are many examples of broadly cultured scientists and engineers who derived technical inspiration from humanistic or artistic thought.

I speak from conviction and a lifetime of experience, and I believe that what I say makes sense and should be convincing. But one can go just so far on reasoned argument. Let us move, then, from deduction to the inductive mode, from logic to observation. Let us look at the world and at the real people in it to see if we can add flesh and blood to the ideas I have set forth.

19

The Civilized Engineer: (2) The People

When I consider the relationship of technical thought to art and general culture, I think first of the nuclear physicists who have done so much to change our world, conceptually and tangibly, by their unlocking of the atom. I think of Albert Einstein at the age of fifteen reading Spinoza, as well as Euclid and Newton, and immersing himself in Beethoven and Mozart. All his life this great scientific genius read history, biography, and philosophical essays, wrote beautiful, contemplative prose, and played the violin. I think of Neils Bohr, who said that he arrived at the concept of complementarity by speculating on the ancient theological dilemma of the impossibility of reconciling perfect love with perfect justice. Of Max Born it was written that "his talents were so various that he might well have become a first-rate pianist or author." Werner Heisenberg read Plato and was thrilled by the atomic theories of the ancient Greeks. According to a friend, he climbed on the red cliffs of the island of Heligoland reading Goethe and "in the intervals" developed his ideas on quantum mechanics "in a kind of intellectual intoxication." Wolfgang Pauli was said to have conceived of the exclusion principle while attending a musical revue, although in later years he maintained that the idea came to him "during a simple promenade."[1]

One sees them all in the mind's eye, promenading along the streets of ancient university towns or on long walks in the countryside, quoting poetry, debating philosophy, and at the same time working out mathematical and physical concepts of sublime originality. Edward Teller is known to us today as the somewhat forbidding "father of the H-Bomb," but when he studied with Bohr in Copenhagen in the early 1930s he wrote poems in secret and was known for his love of literature and philosophical speculation.[2]

Nor was the tradition of broad interdisciplinary culture limited to Europeans. J. Robert Oppenheimer studied Dante, was fond of quoting Proust, and read the works of Indian sages in the original Sanskrit. He applied his eclectic intellect not only to solving problems in physics but also to a spectacular career in government and academe. One of the scientists whom he recruited to work at Los Alamos commented on his "intellectual sex appeal."[3] His eventual downfall—related to associations with people of radical political views—stemmed, like his meteoric rise, from interests that far transcended pure science. Even his greatest detractors admit that in overseeing the development of the atomic bomb his achievement was extraordinary.

I. I. Rabi, who won the Nobel Prize for physics in 1944 and who was one of the prime movers in the development of radar, relates that as a child he became enamored of history. In high school he scored the highest grade in the subject ever recorded for the New York State Regents Examination. In his late seventies Rabi remarked, "I still have a great interest in history and in its interrelatedness." He also recalled that when he worked on his doctoral dissertation at Columbia he taught twenty-five hours a week, day and night, and for relaxation went to the opera—standing room, of course.[4]

What giants they seem, those heroes of the physics revolution—intellectual, cultural, scientific, and many of them moral and political as well. How naturally they combined liberal learning with their professional specialty. Nuclear physics, of course, is not engineering, but the way in which their lives joined mathematics and science synergistically with philosophy and art gives all of us in the science/technology continuum much to ponder.

* * *

A researcher who recently studied the notebooks, diaries, and correspondence of a number of leading biologists and chemists of the nineteenth and early twentieth centuries concluded that the most creative scientists are likely also to be talented artists, musicians, and craftsmen: "They learn to look more carefully at underlying structures. They are more apt to see analogies, to have better hand-eye coordination."[5]

The relationship of the fine arts to science and technology is at once obvious and subtle. Samuel Morse was a well-known painter and president of the National Academy of Design long before he became interested in telegraphy. As noted by Brooke Hindle, author of *Emulation and Invention,* "The arts of design, as Morse defined them, rest most heavily upon spatial, visual, holistic thinking."[6] Robert Fulton studied painting in London under Benjamin West prior to his involvement with steamboats. Benjamin Latrobe, one of the premier engineers in early nineteenth-century America, was an accomplished watercolorist. John Edson Sweet, a founding member of the American Society of Mechanical Engineers and a noted designer of steam engines, lathes, and other machines, was an ardent admirer of art and architecture. According to his biographer, "Out of his inherent enthusiasm for beauty of proportion grew his characteristic ideas of proportion in machines."[7] The absence of art, or even an interest in art, among engineers of today, coupled with the disappearance of drafting from many engineering curricula, shortsightedly ignores the importance of spatial conceptualization in engineering creativity.

Music, too, has a time-honored association with science and engineering. Indeed, the ancients regarded music as a science of measurement, to be classified with geometry, astronomy, and arithmetic. Musical order was considered a reflection of the harmony of the heavens and a bridge to the harmony of the soul.

Einstein the violinist we have already mentioned. Max Planck, another brilliant physicist, had a lifelong interest in music, and in particular studied the untempered "natural" scale. Francis

William Aston, inventor of the mass spectograph and Nobel Laureate in chemistry, was an excellent musician who became music critic for the *Cambridge Review.* Among engineers, John Rennie, famous British builder of canals, was proficient on many instruments, among them the bagpipes. The Roeblings were both accomplished musicians—the father on the flute, the son on the violin.

Art and music are tied to science and technology by shapes and numbers. With poetry, philosophy, history, and literature the connection is not as apparent. Nevertheless, the humanities inspire and expand the technical imagination in many ways. Among engineers, Nikola Tesla provides a most dramatic example. While trying to develop an electric motor that would not require brushes and commutator (which was prone to sparking and burning out) a flash of insight came, he recalled, as he recited a passage from Goethe's *Faust* while walking with a friend in the Budapest city park:

> *The sun sinks; the day is done.*
> *The heavenly orb hastens to nurture life elsewhere.*
> *Alas, no wings lift me from earth.*
> *To strive always to follow!*
>
> *Oh, that spiritual wings soaring so easily*
> *Had companions to lift me bodily from earth.*

"In an instant I saw it all," says Tesla in his *Personal Recollections.* What he saw was a machine rotating in a moving magnetic field, a vision that led to his invention of the polyphase system.[8]

Such moments of technical inspiration stemming directly from poetic imagery are undoubtedly rare, but time and again we find in the lives of great scientists and engineers a rich association with literature and the arts. Charles Steinmetz of General Electric was a student of the classics and a strong advocate of liberal education for engineers. David Steinman, famed builder of bridges, studied the humanities and three foreign languages, earning a Master of Arts along with his engineering degrees. Even in the unlikeliest places one comes upon evidence of the value of the liberal arts in engineering and invention. According to standard biographical sketches, Thomas Edison was "unschooled," a natural genius who combined inventive brilliance with entrepreneurial acumen. But upon looking a little deeper one learns that his mother took him out of school and tutored him at home because his teacher objected to his "dreaming." By the time he was twelve he had read *The Decline and Fall of the Roman Empire* and Burton's *Anatomy of Melancholy.* He became a voracious reader with the gift of total recall, and in his mature years he read widely, listing Shakespeare and Tom Paine among his favorites.[9]

More important than any specific connection between liberal culture and technical creativity is the overall effect on character and personality that can be observed in those engineers whose education was liberally enriched and whose interests were not limited to their everyday work. The Roeblings are an impressive example.

John Roebling, before coming to the United States, studied architecture, bridge construction, and hydraulics at the Polytechnic Institute in Berlin. He also studied philosophy under the great Hegel, and throughout his life wrote essays reflecting profound spiritual concerns. Washington Roebling, who was placed in charge of the Brooklyn Bridge project when his father died in 1869 just as construction was about to commence, had similarly broad interests and talents. He had attended R.P.I., where, in addition to the many required technical courses, he studied Logical and Rhetorical Criticism, French Composition and Literature, and Intellectual and Ethical Philosophy. He was fluent in French and German and had a lifelong taste for the works of Schiller, Goethe, Carlyle, Thackeray, and his favorite, Tolstoy.

He played chess, went often to the opera, and was, as noted earlier, an accomplished musician. It is difficult for us today to appreciate the esteem, approaching awe, with which these engineers were regarded. Their ability to persuade, enlighten, and inspire their fellow citizens contributed as much to the success of the Brooklyn Bridge project as did their considerable technical ability.[10]

Personality and character play a crucial—and fascinating—role in the success of many engineering projects, and liberal education has historically contributed to that intangible that we might call "engineering charisma." Among the leaders in designing and building the first electric power systems in Britain were the Merz family of Newcastle. J. Theodore Merz, after studying chemical engineering, founded the Newcastle Electric Supply Company and served as director of a leading electrical manufacturer. He was also the author of one of the major scholarly studies of his time, the four-volume *History of European Thought in the Nineteenth Century.* His son, Charles, after an apprenticeship in power plants and engine factories, founded a consulting engineering firm and later became a member of the parliamentary committees that planned the reorganization of the electric supply system in England after World War I. He was famed for his charm and eloquence, and according to historian Thomas P. Hughes, "He was believed by many to be the most effective expert witness in the engineering world."[11] In Charles Merz, intellectual, social, and political factors combined with technical and economic knowledge to create both a successful career and, for the nation, enormous technological benefit.

At about the same time, Oskar von Miller, the leading German consulting engineer, was promoting vast hydroelectric projects, making good use of his powers of persuasion as well as his family's prominent position in Bavarian society. He was named to the *Reichsrat,* or Upper House, of the German Parliament in 1909, but this did not prevent him from carrying on as engineer and entrepreneur. He helped form alliances between government and industry that brought several major works to fruition. Thomas P. Hughes has commented on his "forensic gifts" and "personal magnetism" and noted his lasting influence:

> Even today, for many Bavarians he still symbolizes traditional values such as geniality *(gmutlichkeit),* family loyalty, and love of the land. His speeches and debates in the *Reichsrat* are models of the clear exposition of technical subjects. The record suggests that he himself remains a model of the politically sensitive and effective engineer. . . .[12]

Merz and Von Miller may be the models, but where are we to find their American counterparts? In Admiral Hyman Rickover, perhaps, the widely acknowledged father of the nuclear navy who made his mark, not by using Bavarian charm, but by intellectual brilliance, forceful personality, and political savvy—and who, at the age of eighty-four, said that "the best engineers are those who, in addition to technical expertise, have had good training in the liberal arts and understand the world around them."[13] I also think of Vannevar Bush, the craggy, pipe-smoking Yankee who, while professor and dean at M.I.T. in the thirties, designed the differential analyzer, one of the earliest computers, then went on to become president of the Carnegie Institution, director of the U.S. Office of Scientific Research and Development during World War II, and adviser to presidents. Near the end of his long and illustrious career Dr. Bush wrote,

> There is a pressure on the teachers in any professional school to concentrate on the specific skills which their students will need for their professional careers, and thus to neglect almost everything that would increase their ability and desire to take part in those associative relations with their fellow citi-

> zens which are an equally important aspect of a full life. . . .
>
> Thus we need a balance. Alongside the course in the mathematics of electric circuits we need a course in the history of ideas. And we need that balance wherever older minds seek to help younger minds on the way of life.[14]

It may be said that I have been referring to people who lived in an earlier age, before there was television to watch, or travel by jet plane, before skiing, golf, and tennis gained broad popularity, when hardly anybody went to college and those few who *were* educated were broadly educated because that was the tradition and specialties were not yet fully developed and all-consuming. I recognize that times change, and I do not expect that today's average engineer will play the violin and spend leisure hours reading Dante. I also recognize that one source of engineering lies in simple craftsmanship, and considering this, any higher education at all can be considered an improvement. But if eminent scientists and engineers have in fact evolved from a classical tradition, I believe we should think twice before abandoning that tradition.

Happily, the tradition still lives. There is evidence that some of the most imaginative engineers continue to find nourishment in the world of the liberal arts, particularly so in the vanguard of electronics and computer engineering.

Tom West, head of the team of computer-building engineers whose deeds are chronicled in Tracy Kidder's *The Soul of a New Machine,* is described as a philosophically minded guitar player who took up engineering after studying the liberal arts at Amherst College. Mitchell Marcus, a leading computer scientist at Bell Labs, recalls spending his senior year in high school writing plays, poetry, and song lyrics. He entered Harvard determined to major in philosophy and started out studying Greek ethics. While taking a course in logic he became interested in the philosophy of language and ended up majoring in linguistics. Along the way he studied computers and got to Bell Labs via the Artificial Intelligence Laboratory at M.I.T.[15]

Peter Hagelstein, who at the age of twenty-four developed a laser device that has become a key component in the Star Wars program, played violin in the M.I.T. symphony orchestra and likes to read French literature in the original. Concerned about the moral implications of nuclear weaponry, he turned to the works of Aleksandr Solzhenitsyn. Previously, in high school and college, he had read much other Russian literature and history. In his spare time he composes music. (While this book was in its final stages of editing, I learned that Hagelstein had decided to quit weapons research and devote himself to work that, according to friends, "will benefit all mankind.")[16]

Steven Jobs, fabled co-founder of Apple Computer, Inc., recalls that between his sophomore and junior years at Reed College in Oregon he discovered Shakespeare, Dylan Thomas, "and all that classic stuff." He read *Moby Dick* and started to take creative writing courses. For a time he was interested in Eastern mysticism.[17] Kenneth Olsen, the engineer who founded Digital Equipment Corp. and who *Fortune* calls America's most successful entrepreneur, "thinks about morality and religion far more frequently than about microcircuits or finance."[18]

Alan Kay, who led a computer-designing team at Xerox, became chief scientist at Atari, and then took a top research post at Apple, likes to refer to himself as "a failed musician." After earning a bachelor of arts degree at the University of Colorado he was accepted for graduate work in computers at the University of Utah because, according to the head of the program, "he had done interesting things in theater and television and was a musician. I thought he'd be an interesting person to have around the lab." Kay says that in hiring computer people he looks for creative talent first, technical proficiency second: "I find it much easier to take someone from architecture or the liberal arts and train them in the technical stuff, than to take someone with technical expertise and train them to think creatively. Too much engineering training makes people anal and rigid."[19] An extreme view, perhaps, but it gives one pause.

This handful of examples demonstrates that along the frontiers of high tech, as among the pioneers of nuclear physics, multiple talents and interests are an asset. In other fields of engineering as well, there are examples of people with broad interests and wide-ranging intelligence. At every awards ceremony of Tau Beta Pi, the engineering honor society, one learns of young engineers who paint, perform music, write poetry, and talk politics. When the society was founded in 1885 the preamble to its constitution spoke of "distinguished scholarship and exemplary character," of both "attainments in the field of engineering" and a "spirit of liberal culture." This spirit is kept alive not only by Tau Beta Pi prizewinners, but by engineering students across the land, bright young men and women who have a serious commitment to the humanities and the arts, who refuse to be limited to the "constricted interests" found in the stereotypical engineering profile. They are a relatively small group—let me hazard a guess based on personal experience and say no more than 10 percent of the engineering student body—but they testify to the resistance of the human spirit to overspecialization.

And what does this demonstrate, one might ask, other than—as historian David McCullough once said in an address to the American Society of Civil Engineers—that "engineers are people"? Quoting a letter in which Theodore Roosevelt expressed astonishment at encountering an engineer with refined literary taste, McCullough remarked, "So: isn't this wonderful? The chief executive of the land has found himself an engineer who actually reads books!" Then he continued:

> Engineers who read, who paint, who grow roses and collect fossils and write poetry, who fall asleep in lectures, very human-like, even civilized civil engi-

> neers are scattered all through the historical record. Civil engineers have been known to go to the theater, yes indeed; they have taken pleasure in good music and fine wine and the company of charming women. There is even historical evidence of the existence among a few civil engineers of a sense of humor.[20]

Yes, to be sure, the humanistic tradition in engineering does exist. But the point is that this tradition is under attack—has always been under attack—not by people of evil intent but by circumstances, by the ever-increasing technical scope of engineering work, by limits of time and energy, by historical differences and antagonisms between some engineers and classical scholars, and by the failure of many people to recognize the importance of liberal learning to engineering excellence. The attack is particularly strong in these days of waning idealism and heightened interest in career security.

This attack must be resisted and a counterattack launched. The nation needs engineers who are able to communicate, who are prepared for leadership roles, who are sensitive to the worthy objectives of our civilization and the place of technology within it, and whose creative imaginations are nourished from the richest possible sources—spiritual, intellectual, and artistic. Furthermore, engineers as a group need to preserve their professional self-esteem—and the esteem of the greater community—by guarding against an insensitive mechanical approach to the work they do. And finally, individual engineers deserve the chance to enrich their lives with art, literature, history—the best our civilization has to offer.

We cannot expect all engineers to become at once eloquent, commanding, political, sensitive, and creative. A range of talent is what is needed; in variety there is vigor. But just as we want all citizens to be literate and to gain as much schooling as possible, so must we want for all engineers the greatest practicable exposure to the liberal arts. It would be grand if there could be an instantaneous profession-wide consciousness-raising, with each engineer enlightened, uplifted, and determined to improve. Nothing is more heartening than to come across people in mid-career who exemplify the good qualities implicit in the term "civilized engineer." These admirable individuals make their presence felt in their private and professional lives and in the activities of the many professional societies.

Yet general exhortations to the profession, however worthy, can have only limited results. To be at all effective, our aspirations must be formalized in the education we offer to prospective engineers, and it is this topic that I wish to consider next.

The structuring of educational programs sounds tedious, yet nothing could be more important, particularly when we consider that in the word *curriculum* we encapsulate our dreams for the future.

20

The Civilized Engineer: (3) The Schooling

One tends to take for granted the existence of educational criteria in all professions. Thus, I was surprised to discover that engineering accreditation was not formally established in the United States until 1932, when the leading engineering societies created the Engineering Council for Professional Development (ECPD)—renamed ABET in 1980. Although to most engineers it remains obscure, this organization has been a central force in engineering education, and hence the profession as a whole.

ECPD initially developed minimum criteria for undergraduate engineering degree programs, and in 1936 implemented the evaluation of existing programs and the awarding of accreditation. In 1946 the procedure was extended to cover programs in engineering technology. (The criteria are established by ABET's board of directors, and the accreditation is carried out through direct on-site visitation by designated teams.) Today ABET is governed by nineteen "participating bodies"—the major technical and professional engineering societies in the nation—whose representation on the board of directors is roughly proportional to the number of their members. The U. S. Department of Education formally recognizes ABET's exclusive jurisdiction for accreditation, and state licensing authorities accept ABET-accredited engineering programs as satisfying the educational element of licensure requirements.

As noted earlier, engineering education in America began under circumstances that differed markedly from those of the other leading professions. Schools of medicine and law, for example, were established by practitioners and then somewhat loosely affiliated with universities. Sponsorship for engineering schools came from the government (the U. S. Military Academy in 1802 and the land-grant colleges starting in 1863), from public-spirited citizens who founded technical institutes (for example, R.P.I. and M.I.T.), and from within universities, where engineering departments seem to have been created spontaneously out of thin air. Practicing engineers and educators of engineers were—and to a certain extent still are—different breeds.

Engineering education did not grow out of the apprenticeship system, but rather in competition with it. Prior to the establishment of academic programs, training institutions already existed: they were located on canal construction projects where underlings such as surveying assistants learned on the job, progressing through the ranks until their mentors deemed them worthy of the title "engineer." This tradition was continued by the builders of railroads, and carried on in factories and machine shops long after college engineering programs were established. As late as 1916 half of America's engineers had never been to college at all, and debate between representatives of the "shop culture" and the "school culture" was often intense.

Since academic engineering education was from the beginning undergraduate education, the idea of getting a college degree *before* going to professional school never took hold. College and professional school were the same institution. So it is the four-year bachelor's degree that lies at the heart of ABET's accreditation activities. The so-called 3-2 program exists at Dartmouth and is voluntarily taken by a number of other engineering students, most of whom attend liberal arts colleges for three years and then transfer to cooperating engineering schools for a final two. But this pattern—which usually results in a B.A. degree from the first institution and a B.S. in engineering from the second—accounts for only about 5 percent of all engineering graduates.

From its earliest days, ABET's predecessor, ECPD, was occupied with curricular battles over science versus hands-on engineering, research versus practice, theory versus design, and specialization versus broad basics, and so had little time and energy to devote to the liberal arts. Even when, in the early 1940s, a report from the SPEE (now ASEE) recommended a minimum 20 percent humanistic-social component in the engineering curriculum, ECPD took no action. Not until 1955 did ECPD issue criteria covering the liberal arts, and then the minimum requirement was not the 20 percent that had been called for but rather 12½ percent, the equivalent of a half year out of four. Even this was greeted by cries of protest from non-complying institutions, and ECPD had to issue assurances that this was merely a "statement of principle" and would not lead to withdrawal of accreditation. By 1971 recommendation had become requirement, and today all accredited engineering curricula contain at least a 12½ percent liberal arts component. Some schools—notably the Ivy League, M.I.T., Stanford, and other private institutions—require 20 percent or more. But at most of the large universities, where the vast majority of American engineers are trained, the minimum becomes the maximum and the so-called "ABET requirement" is viewed as a hurdle to be overcome en route to the accredited degree.

William E. Wickenden, in his influential *Report of the Investigation of Engineering Education 1923–1929,* noted that in humanistic studies "the primary criterion is not one of intrinsic cultural or intellectual values, nor one of narrow utility, but that of functional relation to engineering."[1] The "primary criterion" remains essentially the same today, and by stressing such "functional relation," engineering educators have reduced the likelihood that their students will ever become acquainted with Greek drama, Roman history, or the English novel.

Perhaps we should take a moment at this point to discuss terminology. Literally, arts that are "liberal" are those studies deemed fit for liberated, or free, citizens. In the universities of the Middle Ages there were established seven branches of liberal learning: grammar, logic, and rhetoric (i.e., literature and philosophy), and arithmetic, geometry, astronomy, and music (science and the fine arts). Through the years a liberal arts education has come to comprise all those studies that are not technical or vocational by nature, those that are not "useful arts." It is no longer said that the liberal arts are *for* free men. Instead one hears it said that the liberal arts *make* men free by liberating their minds from ignorance and dogma.

A proper liberal education includes the study of science and mathematics, and in recent years some attention to the basics of technology. Liberal learning also embraces the social sciences: sociology, anthropology, psychology, political science, economics, and the like. Finally, there are languages, the fine arts, and music, and then history, literature, and philosophy—the so-called "humanities," which are the true core of liberal learning.

"Skill" courses such as accounting do not count toward the ABET 12½ percent liberal requirement. But there is constant pressure on behalf of utility and against the "pure" humanities, a bias that was reflected in 1984 when ABET first permitted introductory language study to count toward satisfying the humanities component. A foreign language might come in handy on assignment abroad. Of what use is a novel, a poem, or a drama?

Liberal education is not something that any thinking person today is likely to be *against.* The key problem is how much of it can and should be offered to students whose careers require them to learn tremendous amounts of purely technical material. (There has been some talk of teaching technical material in a liberal way, but such a concept obviously has its limitations.)

This question was very much on the minds of the people gathered in Phoenix in the fall of 1985 for ABET's annual meeting. This was evidenced by the cast of invited speakers, which included the executive director of the Council for Liberal Learning of the Association of American Colleges, the Dean of Arts and Sciences of Washington University (whose presentation was titled "The Liberal Spirit and Professional Education"), and me. Appeals for liberalized engineering education were also made by employers of engineers who participated in the Industry/Government Symposium sponsored by ABET the day before the annual meeting.

But when the oratory dies down and all the fine sentiments have been expressed, the situation remains unchanged. The ABET-accredited entry-level degree for engineers continues to be the four-year baccalaureate with a 12½ percent liberal arts requirement. Not only has there been no change, but the pressure for change has engendered resentment. For example, at the ABET meeting the dean of engineering of the University of South Carolina vigorously attacked several recent reports that have called for an increased liberal component in professional curricula. He made the point that engineering educators do more to expose their students to the liberal arts than humanities people do to expose their students to technology. I have seen similar remarks in professional journals attributed to the executive director of the American Society for Engineering Education and to the chairman of the Engineering Deans Council. This response (which is not as valid as it once might have been, and is also, I believe, beside the point) indicates how touchy the situation has become.

One can appreciate the exasperation felt by deans who have to deal with everyday administrative pressures far removed from well-intentioned theorizing. Any interest in liberalized technical education runs head-on into intensified demands for utilitarian vocational training. Most of the young engineering students I meet these days want to take technical courses that will maximize their chances in the job market. They don't want to spend a day or a dollar more than necessary in pursuing their degrees, and they are not at all convinced that the humanities are relevant to their concerns. Corporate employers of engineers, for all their proclaimed interest in the liberal arts, are content to hire the current crop of graduates and to use many of them in sub-professional jobs.

On the campuses, long-standing prejudices and academic politics make change difficult and interdisciplinary cooperation a low priority. It took a long time and many visits to academe before I recognized the nature of the problem, although once understood it seems downright obvious. In colleges and universities with strong, independent liberal arts departments, the best faculty are mostly concerned with teaching, writing, and research within their own specialty. There is little satisfaction, reputation, or advance toward tenure associated with teaching literature or history to engineers. There is also a provincial competition between departments that militates against the broad overview that might be especially helpful to engineering students. In engineering schools where a small liberal arts faculty constitutes a designated resource for the engineering students, there is likely to be more dedication to the job at hand, but this is offset by other disadvantages. The distinction of large liberal arts departments is missing. The non-engineering faculty are often viewed as second-class citizens, and their offerings are thus tainted.

In both sorts of institutions the engineering faculty have considerable influence over the way engineering students perceive their liberal studies, and too often, I regret to say, this influence is not used constructively. Disdain for the liberal arts has become endemic in the engineering community, and the carriers of this disease are too often found among engineering academics. Even fifty years ago the dean of Columbia's school of engineering could observe that "you cannot in general encourage and develop in engineering students interests and viewpoints more liberal and broader than those of the instructors with whom they have their most intimate and constant contacts."[2] Eugene Ferguson, an engineering professor who became an eminent historian of technology, sees little hope for improvement of the situation short of his proposed method, "which is to kill off a generation of teachers and, I suppose, deans, too."[3]

Of course, as a number of engineering professors have told me, they view it as their duty to produce, not Renaissance men and women, but the large numbers of competent engineers that have traditionally made our nation's technology the envy of the world. If American innovation, quality control, and productivity have declined, most engineering educators do not see the humanities as the source of potential improvement.

It is small wonder that long-time observers of the scene tend to become discouraged or cynical. On almost every engineering campus there is a tale to be told of efforts in liberal education that came to naught. On a nation-wide basis, perhaps the most disheartening failure was an ill-fated enterprise sponsored by the Sloan Foundation.

In December 1971, the Alfred P. Sloan Foundation sponsored a new program called "The Social Dimensions of Engineering Practice." The attitude of Americans toward technology had been souring for a number of years, and with the onset of the "environmental crisis" in 1970, engineers were everywhere on the defensive. The times seemed to require engineers who could understand the cultural and social implications of their work and who could cope with the rapidly changing political and economic environment in which they would have to make their careers.

The Sloan trustees confronted the question of how to educate such engineers and decided upon a new and imaginative approach. They invited a number of the nation's leading engineering schools to apply for grants to support experimental interdisciplinary activities. The "Social Dimensions" program ran from 1972 to 1977, during which time the foundation gave almost $9 million to 23 institutions. The money was used mostly to create new courses, more than 200 in number, with titles such as "Technology and Human Values," "Interaction of Public Policy with Technology," and "Environmental Issues and Problems." In this endeavor, hundreds of engineering faculty members worked jointly with colleagues from the social sciences, and to a lesser extent from law and the humanities.

The survival rate of the new courses, according to the summary Sloan report, was "not very high," and as for the interdisciplinary teachers, most "returned to their regular assignments." Still, Sloan deemed the program to have been valuable in inspiring interaction between engineering and the other academic disciplines, and in helping to introduce technology into the mainstream of liberal education.[4]

Certainly the effort was worth making, however ephemeral the results. There was one aspect of the program, however, that the Sloan people found "surprising and disappointing": *the newly developed courses appealed mostly to non-engineering students.* In other words, "The Social Dimensions of Engineering Practice," instead of improving the liberal arts education of engineers as intended, revealed an interest in technology among students in the arts and sciences. This discovery must have influenced the foundation officers and staff as they considered their next major program, "The New Liberal Arts." In March of 1982, a number of select liberal arts colleges were invited to apply for grants aimed at integrating "applied mathematics and technological literacy" into their curricula. According to the letter of invitation: "Any attempt to meet the purposes of the program will almost certainly entail instruction intended to create 'computer literacy.' "[5] Within three years the foundation invested more than $10 million in a program involving more than fifty colleges and universities.

The turnabout was breathtaking. Having given up on the idea of creating humanistically sensitized technologists, the foundation had decided to help form technologically enlightened humanists. The successes of the second program, while gratifying in their own right, add poignancy to the failure of the first.

* * *

Despite all the discouraging signs, it is clear that the ideal of liberal education for engineers will not be abandoned. Its continuing presence on the agenda of engineering society meetings indicates that the idea has considerable vitality. National leaders, recognizing the importance of engineering education to national well-being, are investing money—the surest sign of concern—in pursuit of improvements. The Association of American Colleges, with a grant from the Andrew W. Mellon Foundation, embarked in 1986 on a project "to promote new patterns of engineering education," with special emphasis upon strengthening the quality and coherence of the humanities and social science elements of the curriculum and with the aim of preparing students more effectively for management and leadership responsibilities. The National Research Council study, "Engineering Education and Practice in the United States" referred to earlier, grappled with the liberal arts problem but found no ready solution:

> We are aware of intense pressures to modify the undergraduate engineering curriculum to include more subjects in the humanities, liberal arts, and social sciences as well as more technical and business courses, all within the confines of a sacrosanct four-year program. Arguments on all sides are unimpeachable but they are also mutually exclusive, and moving in favor of any one of them causes the root curriculum to suffer. The arguments could be reconciled in a plan for a pre-engineering undergraduate program followed by a professional school program, with the combination requiring more time to earn the first professional degree. However, because of objections to the extra costs of this approach and the expected reluctance on the part of students to extend their college program, the committee could not reach a consensus on this vexing problem.[6]

Vexing, indeed.

Rather than let the matter rest, however, the study committee recommended that the National Science Foundation fund a pilot group of engineering schools to evaluate alternative programs experimentally.

A realistic appraisal of the current scene—taking into account the attitude of students, the mood of the Engineering Deans Council, and the expedient needs of the nation—leads one to conclude that change in the minimum accreditation requirements will not come soon. Certainly a universal shift to a five-year program is not politically feasible. "Utopia," says the dean of engineering at Ohio State, "but it just won't work." "An attractive consideration," according to the dean at Iowa State, "but not a very practical one." Even among engineering educators at smaller liberal arts institutions the idea presents difficulties. The dean of engineering at the University of Pennsylvania cautions that adding a fifth year could mean the end of engineering at such schools. "If we get too opinionated about professionalism and add a fifth-year requirement," he warns, "we're going to kill off the diversity offered by these kinds of institutions."[7]

As for increasing the minimum liberal arts requirements, this also would be fought tooth and nail. At institutions where the stress is utilitarian, and students are prepared for entry into defined jobs in particular industries, the deletion of even one technical course is considered intolerable. And even at schools that are grappling with ways to enrich their liberal program, additional direction from ABET may not be welcome. Accreditation standards, we must remember, set mutually agreed *minimum* requirements. When these standards become overly precise, they are liable to frustrate creative individuality in the structuring of curricula. Superior academic institutions, at every level and in every field, are constantly on guard against the stifling, homogenizing influence that can come in the guise of accreditation demands.

Accordingly, the next step in liberalizing engineering education will not—and should not—come from changes decreed by ABET, but rather from the soul-searching of individual institutions. And indeed, as announced in a front-page article of *The New York Times* in 1985, "a new wave of curriculum reform" is sweeping through American colleges. Following in the wake of Harvard's much heralded Core Curriculum (approved by the faculty in 1978 and fully implemented with the junior class in 1984–85), hundreds of institutions have taken steps to restore traditional "distribution requirements." "What we have done, in essence," says the president of Skidmore College, "is to redefine our concept of what constitutes an educated person."[8] This nationwide movement, together with the manifest uneasiness of engineers about creeping vocationalism, is likely to have a considerable effect in engineering education.

The planning of a curriculum—particularly for those of us who do not have to deal with academic departments, industry demands, and student pressures (to say nothing of families, alumni, university boards, and state legislatures!)—can be a pleasant exercise much like choosing what books to take along to a desert island. The ideal engineering curriculum, we can all agree, would start with a thorough grounding in the humanities—along with mathematics, theoretical science, and social sciences. It would provide a broad spectrum of basic engineering knowledge, good for a lifetime of ever-varying challenges, along with precise skills suitable for productive entry into the workplace. In a perfect world, engineering students would be taught creativity—assuming it can be taught. They would learn how to draw so that they could think with their hands as well as their minds. They would learn economics and develop business sense. They would learn how to do research and how to run a factory. They would know all about materials and electricity, all about light, heat, fluids, even biology and the human body. They would study psychology, ecology, and safety engineering. They would learn all of this in breadth and also in depth. They would know how to write clearly and speak eloquently. They would develop stalwart characters and be well versed in politics and professional ethics. They would study together with their fellow engineers and also mingle in classes with students of other disciplines. They would acquire all the best features of Leonardo, Einstein, Edison, Madame Curie, and . . . well, you get my point: We have no choice but to stop dreaming, wake up, and deal with the world at hand.

Because we cannot do everything we want to do, we are forced to make choices. In a hospital emergency room this unavoidable decision-making is called *triage.* All things considered, what should we do and what must we reluctantly leave undone?

On the technical side there seems to be widespread agreement that what is wanted is a broad foundation in the basic engineering sciences. Past a certain point, specialization must be left to graduate school or training in industry. In the same spirit, I would opt for breadth in the liberal arts.

First, of course, must come the ability to handle the English language. But the goal must be more than the ability to read and write. What we seek is cultural literacy. E. D. Hirsch, Jr., a professor of English at the University of Virginia, has noted that cultural *illiteracy* begins in our elementary and high schools, where many of the students are not getting the ABCs of knowledge. This "knowledge"—our shared cultural heritage and the language of our public discourse—consists of people: John Adams, Benedict Arnold, Daniel Boone, John Brown, Aaron Burr, John C. Calhoun, Henry Clay, James Fenimore Cooper, Lord Cornwallis, Davy Crockett, Emily Dickinson, Stephen A. Douglas, Frederick Douglass, Jonathan Edwards, Ralph Waldo Emerson, Benjamin Franklin, Robert Fulton, Ulysses S. Grant, Alexander Hamilton. It consists of water and mountains: the Antarctic Ocean, Arctic Ocean, Atlantic Ocean, Baltic Sea, Black Sea, Caribbean Sea, Gulf of Mexico, North Sea, Pacific Ocean, Red Sea; the Alps, Appalachians, Himalayas, Rocky Mountains, Mt. Everest, Mt. Vesuvius, the Matterhorn. It consists of our mythic heritage: Adam and Eve, Cain and Abel, Noah and the Flood, David and Goliath, the Twenty-third Psalm, Humpty-Dumpty, Jack Sprat, Jack and Jill, Little Jack Horner, Cinderella, Jack and the Beanstalk, "The Night Before Christmas," Peter Pan, Pinocchio, The Princess and the Pea. It consists of patriotic songs: "The Battle Hymn of the Republic"; "Columbia, the Gem of the Ocean"; "My Country, 'Tis of Thee"; "America the Beautiful"; "The Star-Spangled Banner"; "This Land is Your Land"; "Yankee Doodle"; and so forth.[9]

And just as it is important for children to learn the ABCs of knowledge, so is it vital for college-educated citizens to delve deeper, to learn more of the cultural alphabet. The "Report on the Humanities in Higher Education," issued in 1984 by the National Endowment for the Humanities and authored by William J. Bennett, subsequently Secretary of Education, contains a further basic list—a list not merely of names and places but of authors whose works should be read, or at least sampled, by those men and women who would partake of higher education: from classical antiquity—Homer, Sophocles, Thucydides, Plato, Aristotle, and Vergil. From medieval, Renaissance, and seventeenth-century Europe—Dante, Chaucer, Machiavelli, Montaigne, Shakespeare, Hobbes, Milton, and Locke; from eighteenth- through twentieth-century Europe—Swift, Rousseau, Austen, Wordsworth, Tocqueville, Dickens, Marx, George Eliot, Dostoyevsky, Tolstoy, Nietzsche, Mann, and T. S. Eliot. From American literature and historical documents—the Declaration of Independence, the Federalist Papers, the Constitution, the Lincoln–Douglas debates, Lincoln's Gettysburg Address and Second Inaugural Address, Martin Luther King, Jr.'s "Letter from the Birmingham Jail" and "I Have a Dream" speech, and such authors as Hawthorne, Melville, Twain, and Faulkner. And, of course, the Bible. And there is a further need for the educated citizen to gain familiarity with the great works of music, art, and architecture.

The humanities, according to Bennett, are not an educational luxury, nor are they just for humanities majors. "They are a body of knowledge and a means of inquiry that convey serious truths, defensible judgments, and significant ideas. Properly taught, the humanities bring together the perennial questions of human life with the greatest works of history, literature, philosophy, and art."[10] (As an important qualification to Bennett's approach—and Hirsch's for children—I would stress the need to encourage individual and institutional variations. Too often, "standard" lists ignore ethnic and social diversity.)

It seems to me that anyone who would call himself college-educated—particularly anyone who would call himself a professional—should spend some time in close communion with the great souls, the great thinkers, the great artists, of our civilization. Most particularly, those engineers who would be leaders, those who would participate in the important communal debates, should be acquainted with the thoughts, theories, and philosophies that constitute the foundations of our culture. Certainly, science and technology should take their proper places among the important ideas and concepts that constitute the core of civilized knowledge. This has only recently begun to happen. But the narrow world view of a few classical scholars should not serve as an excuse for engineers to be contemptuous of liberal learning.

If I were king, every engineering school would require a two-year survey course in history and literature comparable to Columbia University's Humanities 1001 & 2 (Masterpieces of European Literature and Philosophy) and Contemporary Civilization 1001 & 2 (Development of Western Institutions and Social Ideas). As I noted earlier, such courses once were widespread, but many succumbed to changing fashion and to the objection that they had breadth without depth. This fault-finding comes mostly from academics who would rather teach *how to do* history than history itself, would rather dissect one book than read five or—heaven forbid—read excerpts from fifty. Admittedly, the purposes and methods of historical and literary research are well worth studying, but for the engineer who has only a few precious hours available for the humanities, method must not be allowed to take the place of substance. As I said when my turn came to address the ABET meeting in 1985, *there is more depth to be found in a few days with Shakespeare or Thucydides than in four years with the methodology of most academic specialties.* Let us bring back the survey courses and make them required. Let each institution mold them in its own fashion, make them as lively and rich as possible, and put them in the hands of the best available faculty. If we are lucky, such courses will ignite sparks of curiosity and fan flames of intellect among aspiring engineers. At the least, such courses will expose every engineering student to some of "the best which has been thought and said in the world" (the phrase is Matthew Arnold's). And if anyone—student or faculty—dares to call these courses "Mickey Mouse," then perhaps it is time to stand up to such a person and say, "Be careful, my friend, lest you define yourself. You are the one who stands in danger of becoming Mickey Mouse."

Beyond this relatively modest proposal I will not attempt to design a curriculum. I am not a professional educator, and besides, at this point what we need is a variety of approaches. In my 1968 book, *Engineering and the Liberal Arts,* I spoke of "bridges" of interest that would appeal to engineers—for example, history that stresses technological change, or novels with engineer protagonists. Since the mid-1970s there has been a burgeoning academic field called Science, Technology, and Society. Although I favor such attempts to entice engineers and to relate technology to other academic fields, there is also much to be said for taking the humanities "straight" on their own terms. There is no single best path to learning and civilization.

As an example of what can be done, consider Stanford University. A Western Civilization course that was abandoned in 1970, after having been part of the curriculum for thirty-five years, was restored (its name slightly modified to Western Culture) starting with the class of 1984. This year-long course, which must be taken by all undergraduates, is available in a number of tracks: some stress history, others great literature and philosophy. One sequence is entitled, "Western Culture and Technology." Whichever path is chosen, however, the student takes the grand journey from classical antiquity to the modern age, becoming acquainted with the significant cultural eras, the notable leaders and creators, and the important ideas that constitute the heritage of civilized America. There are seven other distribution requirements, three that engineers automatically fulfill (math, science, and technology) and four others in various areas of the humanities, social sciences, and fine arts. In at least one course the emphasis must be on a non-Western culture. There are also two required courses in writing and, in addition, a foreign language requirement (which can usually be satisfied by three years of high school study). In pursuing this program the typical engineering student earns 23 percent of his or her credits in non-technical areas.

Of course, changes made at an elite private university like Stanford, whose engineering school awards only three hundred B.S. degrees each year and whose graduate department (about a thousand masters and Ph.D.s annually) is the largest in the nation, are not necessarily feasible at other institutions such as Texas A&M, Penn State, Illinois, Ohio State, Purdue, and Georgia Tech (the top six in numbers of B.S. degrees awarded), where engineering undergraduate class sizes range from twelve hundred to more than seventeen hundred.

In addition to changes in course offerings, there are many ways in which engineering students can widen their horizons. Seminars, debates, attendance at plays and concerts, experiments in the arts—extra-curricular opportunities are unlimited. An associate dean at Texas A&M has urged young engineers to travel and has himself led engineering students on summer tours abroad, participating in cultural activities and visiting museums as well as engineering installations. If the will is there—if administration, faculty, and students become convinced of the value of the enterprise—there will be no shortage of creative ideas. Even a new attitude among faculty charged with guidance counseling would go a long way toward improving the situation.

Liberal enrichment of the four-year curriculum, no matter how well conceived, is not likely to satisfy proponents of a five-year engineering degree. Out of the currently widespread self-examination, one might hope that some institutions will develop longer compulsory programs, either similar to the Dartmouth 3–2 plan or original in design. Any of a score or more select private institutions could do this today. The current popularity of engineering as a career choice, and the special cachet of a top-rated school, makes it financially feasible for such schools to make the change. It is no longer true, if it ever was, that all institutions must go to a 3–2 program at the same time if the move is to succeed. It would help, surely, if such five-year programs as are devised could culminate in a master's degree. The B.A. plus a B.S. in engineering is not adequate recognition for an engineer who has received a full-fledged liberal education. This truth is not presently recognized at ABET nor will it be until a number of respected institutions devise programs that are worth championing and can muster adequate political support.

It is high time that we disengage from the battle over extending the engineering program and increasing the minimum percentage of liberal content in the curriculum. Rather, the battle should be shifted away from the narrow confines of ABET accreditation to the broad expanses of creative private initiative. It is no longer useful or appropriate to train all engineers in the same way. The profession should acknowledge that a spectrum exists, ranging from solid, useful, technically adept engineers at one end to socially involved, culturally sensitized engineers at the other. The ideal is to have all members of the profession move toward literacy, culture, and humanistic sensibility while at the same time rising to new heights of technical skill and ingenuity. But, as engineers, we pride ourselves on our sense of reality; let us get on with that part of the ideal that is presently within our reach. Initiatives from elite institutions—and from the elite individuals who are present in *all* institutions—constitute the next logical step. Such initiatives, let us hope, will generate attention, interest, and in due course, emulation. When there are successes to point to and when the mood is less confrontational than it is today, then let the matter be brought before ABET for action. In the meantime, ABET can play a constructive role by fostering discussion, disseminating information, and nurturing respect for the liberal arts.

Engineering students themselves may accelerate this process. The National Research Council report, referred to earlier, notes that today's engineering students "have a richer educational and cultural background and are more confident, more assertive than engineering students of years past."[11] Perhaps some of these able young men and women will look at the world around them and decide that the academic path presently laid out before them is not equal to their talents and their aspirations. By their choices—and by their demands—they may help to shape a richer curriculum and a more noble profession.

21

The Civilized Engineer: (4) Women

The evolution of a truly civilized practice of engineering will be hastened by the entry into the profession of an increasing number of women. I hesitate to write about women and engineering, just as I hesitate to write about ethics, because it is a topic charged with emotion, one that has brought down on my head more impassioned comment than I care to recall. When my article "Engineering and the Female Mind" (the title was coined by the editors) appeared in *Harper's* in 1978 it elicited more mail than anything else I have ever written. Many of the letters expressed fury, some of them accord, all of them fervor.

In that article I challenged the reasons usually given for the low numbers of female engineers—male hostility, teacher prejudice, lack of role models, et cetera—and suggested that talented young women were avoiding engineering because they perceived other professions as a more direct route to political power and social prestige. In 1978 the representation of women among freshman engineers was 12 percent and rising. As of 1986 it had leveled off at about 16½ percent (15 percent of the degrees). This is still inordinately low compared to law, medicine, business, and the sciences, and I believe it is related to the engineer's relatively humble social status. The perception is widespread—among minority students as well as women—that if you're smart enough to be an engineer, you're too smart to be an engineer. In other words, for the same talent and effort, other professions yield greater returns. Of course, this view fails to take into account the psychic returns that are so large an element of the satisfaction that engineers find in their work, and for this I take issue with women today just as I did in 1978.

But no matter how limited their numbers, women are bringing to engineering some fresh and valuable elements: a deep interest in the liberal arts and a concern for the humanistic aspects of the profession. This point has been noted by Vladimir Shlapentokh, an emigré Russian sociologist. In the Soviet Union, it seems, women engineers devote three times as much of their leisure time to "humanitarian and artistic activities" as do their male counterparts. They read more fiction and go to museums, theaters, and concerts more often. An unofficial study revealed that they are more willing than men to maintain relations with people frowned upon by the authorities. Asked about the government's treatment of Andrei Sakharov, only 5 percent of the women endorsed the party line; among their male colleagues the figure was 24 percent.[1]

Similar findings have surfaced in the United States. A survey of freshmen at fifteen engineering schools[2] found that three times as many women as men enjoyed reading novels, short stories, drama, and poetry, while the men preferred nonfiction and sports in the same ratio. Women were much more interested than men in participating in preprofessional campus organizations. They were more likely to choose engineering as a career because the "work itself is interesting." More men than women were concerned about anticipated earnings. A study of Purdue engineering students[3] found that women felt more strongly than men that a liberal arts background was essential for engineers. A sociologist at Franklin and Marshall College, interviewing students at twenty accredited engineering schools, found that women were more inclined to enter the "humanitarian" fields such as environmental and biological engineering, while men were more interested in aerospace, electrical, mechanical, nuclear, and petroleum engineering. This analyst also found that women engineering students in general came from a higher socio-economic background than men.[4]

Even in engineering technology colleges, where women are *less* affluent than men, their wider cultural interest pertains. A study at the Wentworth Institute of Technology in Boston revealed that women students came from homes with 20 percent lower income than male students, and in general had "more in common" with their male counterparts than with relatively well-to-do women engineering students. Still, they were twice as interested as the men in art, reading, and the humanities.[5] Thus the difference seems not to be associated with money or social class, but rather with gender.

Although these are just straws in the wind, the findings are remarkably consistent. It appears that women do indeed bring a new outlook to engineering, one that is marked by—in the words of a female philosopher at an engineering school—"greater emotional sensitivity, more extroversion, stronger intellectual interests, and broader concerns beyond the narrowly technical."[6]

According to one view of the female sex this is only what one would expect. Marilyn Ferguson, in *The Aquarian Conspiracy,* puts it this way: "Women are neurologically more flexible than men, and they have had cultural permission to be more intuitive, sensitive, feeling."[7] Sherry Turkle, author of *The Second Self: Computers and the Human Spirit,* has observed that the "entities" that appear on the screen of a computer "stand between the world of physical objects and the world of abstract ideas, and they are taken up differently by hard and soft masters."

> The hard masters treat them more as abstraction, somewhat like Newtonian particles; the soft masters treat them more as dabs of paint, building blocks, cardboard cutouts. You can find examples of hard and soft mastery among both boys and girls. But I have found that girls tend to be soft masters, while the hard masters are overwhelmingly male. . . .
>
> The response of "soft masters" to programming—if they are given the time and the permission to de-

> velop their own way of approaching the machine—has a great deal to teach us.[8]

To some people such associations of women with sensitivity or "softness" is anathema. When asked about the "new femininity . . . the emphasis on feelings rather than intellect," Simone de Beauvoir, whose *The Second Sex* is a landmark in the feminist movement, answered unequivocally: "I think it's a return to the enslavement of women, pure and simple!" As for claiming that women are closer to nature than men: "These are attempts to divert women from their struggle for emancipation and to channel their energies into subsidiary concerns."[9]

Obviously one can't have it both ways. Either women are more sensitive than men or they aren't. The safest course is to say that they aren't. Thus Stephen Jay Gould, reviewing a feminist book on biology, made this cautious statement: "We desperately need more women as equal companions . . . not because the culture of feminism grants deeper vision but because we need as many good scientists as we can get."[10]

I cannot speak of biologists, but as for engineering I will venture where Dr. Gould was reluctant to tread. I believe that women definitely do bring a new dimension to the profession, perhaps not a "deeper vision," but a broader view, a more philosophical and aesthetic concern. I do not suggest that there is anything biological about this—although I do not rule out such a hypothesis—merely that many women come to engineering from a different intellectual background, from a different cultural climate. And even though the doors to engineering have been wide open to females for almost a generation, the young women choosing the profession today are still, by definition, venturesome, that is, willing to go where relatively few of their sex have gone before; whereas for men the profession has become so sensible a career choice as to verge on the pedestrian. An influx of adventurers, hailing from different cultural landscapes and bringing with them a humanistic view of technology, is just what our somewhat insular profession needs.

I hasten to add that this does not mean that women are less adept than men in conventional engineering tasks, and their work in the classroom, at least, bears this out. However, if women sense that engineering is not a profession in which they can readily give expression to their humanistic interests, then perhaps they will *on that account*—as well as for the social reasons previously noted—choose different careers. Those women engineers I have met, or about whom I have heard, do not, in fact, give evidence of unusual sensitivity or broad concern. Rather, they seem to act as much as possible like their male colleagues! So we may be in a catch-22 situation in which expression of female values is constrained because of the small percentage of women in the profession, while other women refuse to enter the profession because they see no evidence there of the values they profess.

It is important to stress that I am not talking about refinement of manners. I heartily approve of women engineers adopting rough-and-ready ways where roughness and readiness are appropriate to the task at hand. Nor am I talking about superficial cultural veneer. I am addressing the need for engineers to be sensitized and liberally educated—so that their own lives can be enriched, their profession ennobled, and society improved.

Women bring to engineering a heightened appreciation of the humanities—and of the values inherent in humanistic education. Studies show this. What the studies do not show—and what nobody can predict—is to what extent this potential influence will manifest itself in real change. The neo-conservative mood of the moment surely inhibits any substantial move toward sensitivity or "soft" mastery. Yet the obstacles to change may provide the very stimulus needed to effect it. A cherished experience in engineering is the phenomenon of *breakthrough.* Perhaps this can happen socially and philosophically, as it has happened so often —and so gloriously—in the technical sphere.

22

The Civilized Engineer: (5) Prospects for Change

Several politicians have been quoted as saying that they would like some day to meet a scientist or engineer who is one-armed —so that he will not be able to say "on the other hand." This rather feeble witticism reflects an all too prevalent impatience with uncertainty. We crave clear answers, but life presents us with complications and inconsistencies. According to a study in *The Harvard Business Review,* one of the most important attributes required for leadership is "tolerance for ambiguity."[1] One of the pleasures of engineering is the non-ambiguous nature of most scientific facts, but one of the greatest challenges of engineering is coming up with a design or an answer when no clear-cut, ideal solution exists. When we turn from engineering problems to people problems, the level of ambiguity rises exponentially.

And so it has been in this extended essay as I have tried to reconcile diverse points of view, all valid yet some of them verging on the incompatible. Should engineers follow their conscience? Of course. Must they subordinate personal views for the sake of organizational efficiency and democratic order? Also of course. Is engineering an arm of industry? Certainly. Is engineering central to the regulatory function? Again certainly. Do we want industrial development or pastoral tranquility? Both. Do we want guns or plowshares? Plowshares, but it isn't that simple. Are female engineers the same as male engineers or are they different? Yes.

Sometimes I feel like quoting Walt Whitman:

Do I contradict myself?
Very well then I contradict myself.

But engineers are not poets and we are reluctant to leave contradictions unresolved. Rather, we like to be clear about those contradictions that we have to live with (differences of personal opinion and philosophy) and those we must resolve by action, compromise, or trade-off.

A topic that I find troublesome, and around which I have been moving in gingerly fashion, is—let me express it in a single word—elitism. The concept of the civilized engineer—the humanistically educated engineer—coupled with the recognition that not all engineers want to, can, or, in any event, will move in the direction I propose, gives rise to the conclusion that we must look, at least initially, to a select group—an elite.

An interesting word, *elite,* one that elicits strong feelings in engineers. There are many who believe it runs counter to the profession's democratic heritage, and also to the practical needs of the nation. In his *Report of the Investigation of Engineering Education 1923–1929,* W. E. Wickenden spoke bluntly:

> There would be no conceivable gain to society in making scientific technology the monopoly of a restricted professional group, as in medicine and law, nor is there any inherent basis for limitations of numbers in technological education.[2]

In other words, the more engineers the better, and if they can do the technical jobs for which they are needed then why bother to screen or "elevate" them?

In 1983 the Chairman of the National Academy of Engineering, in a study entitled *U. S. Scientists and Engineers,* suggested that "it may be necessary for the United States to reduce the current premium placed on elitist behavior. . . ."[3] And a panel participating in the previously cited 1983–1986 National Research Council study warned against "a somewhat elitist rejection of those perceived as not holding the 'proper' credentials."[4]

Yet Jerrier Haddad, the former IBM vice president who chaired the NRC study committee, has suggested that even if present degree programs satisfy "many if not most" engineering students, the ablest students should be encouraged to select a more comprehensive course, five or six years in length, leading to a new but as yet unnamed degree. Haddad proposes that as a start 10 to 15 percent of the engineering student group might avail themselves of such an option.[5]

This is a higher percentage than the one-in-twenty "hotshots" that my previously quoted executive friend wished to have among his engineer employees. It is lower, however, than the percentage that Nobel Laureate Ernest Rutherford, head of the Cavendish Laboratories of Cambridge University, once discussed with Gerard Swope, president of GE. "We devote ourselves to the top quarter of the class," said Rutherford. Swope recalled the conversation in later years:

> I came back filled with the idea that if we could keep the average of our engineering graduates as high as it has been and still give more time and opportunity to the exceptional student, we would make a real contribution to engineering education and eventually to society, by engineering leadership.[6]

Elitism, properly construed, is a profoundly democratic concept. Each person should have the opportunity to rise to the full extent of his or her capacities. Men and women of superior achievement can form centers of leadership without adversely affecting the rights or the dignity of each citizen. Indeed, the recognition of excellence provides pride and inspiration for all. There is, of course, no such thing as a single elite; only a multiplicity of exceptional groups. This is certainly true within engineering, and the various aristocracies often deliver lively and wholesome mutual critiques. I have met outstandingly creative designers and researchers who look down upon the "political" engineers who hold high executive positions in industry or leadership positions in the professional societies—and vice versa. One cannot excel simultaneously in all directions. "For everything you have missed," said Emerson, "you have gained something else; and for everything you gain, you lose something else." The first law of thermodynamics puts in scientific terms a truth that is central to the human condition: the amount of energy in a closed system is constant; it can be changed in form but not in quantity. I am acutely conscious of this myself. During the hours when I write about engineering I cannot be *doing* engineering.

Nevertheless, after all qualifications have been voiced, engineers should be able to agree that the profession must be improved—advanced, enhanced, uplifted—and that a key element of such improvement could be the development of an elite cadre of liberally educated engineers—leaders, communicators, and technician-philosophers. It is difficult to foresee exactly what sort of people they will be, this new breed of engineers. We have models from the past, hints from the present, and visions of the future, but things rarely turn out neatly in accordance with prophecies. Excellence, we know, can take many different forms. It is also difficult to predict where these men and women will come from. The select schools are a likely source, but by no means the only one. I receive letters every so often from engineering students in backwater places whose fierce determination to broaden their education is fueled by the very limitations of their environment. The history of the professions and the arts is replete with individuals who rose to eminence from obscure or isolated beginnings. I also meet deans and professors in conventional institutions who would like nothing better than to be in the forefront of those who will lead engineering into a new golden age.

Often, when engineers discuss the elevation of their profession, someone will raise the subject of nomenclature and titles. If there are to be superior engineers, what shall they be called? The word *ingenieur* has been suggested, in an effort to associate with the prestige accorded professionals in different parts of the world. Other proposed honorifics include ingenior, genior, diplomate, doctor engineer, and master engineer. None of these possibilities seems very promising, and the effort itself verges on the pompous. We would do well to concentrate on more substantial concerns.

The engineering elite of the future will most likely be graduates of liberally enriched five- or six-year programs. But the length of time spent in school is not as important as the quality of that time. Which is to say that some graduates with four-year degrees may attain higher levels of cultural distinction than young people who take extended programs but do not open themselves up to new opportunities. What we are considering is an awakening of mind and spirit.

There are many dull, pedestrian career paths, but the person who follows them, as Charles Steinmetz said to engineers of an earlier age, "does not necessarily have to be *you*":

> There is no law that says once a man gets into a blind alley job, he is doomed for life. The same principle that makes one who has fallen accidentally into a river fight his way to the top to get the blessed air and to preserve his life, should be working within a man as soon as he sees he has fallen into one of these jobs.[7]

As it is with a job, so it is with a career or an attitude toward one's profession.

In fighting their way "to get the blessed air," many engineers start from a discouragingly low level. Consider that two-thirds of American engineers do not bother to obtain a professional state license, or join a single professional society, or take graduate courses beyond the four-year degree—to say nothing of the lack of liberal content in that basic accredited degree. Franklin Roosevelt once said that he saw one-third of a nation ill-housed, ill-clad, and ill-nourished. Engineers who care about their profession must ruefully admit that they see two-thirds of their colleagues unlicensed, unaffiliated, and inadequately educated.

During the course of the recent National Research Council study, members of the panel charged with statistical analysis were amazed to find that there were more engineers in the United States than there were supporting technicians. This seemed to run counter to the practical experience of most participants in technical work. The mystery was solved when the panel realized that large numbers of engineers were working in sub-professional tasks, meaning they are nominally engineers but actually technicians. The finding related to the IEEE surveys referred to earlier, which found that many of the engineers who are dissatisfied with their careers claim they lack opportunities to perform real engineering work.

Yet engineers are intelligent, more intelligent than the average college student, as measured by SAT scores, verbal as well as mathematical. They are, by and large, hard-working, productive, and well-meaning, the very salt of the earth. I reaffirm what I wrote a decade ago in *The Existential Pleasures of Engineering:*

> Intelligent, energetic, unassuming people who seek interesting work! Can this be mediocrity? The very least that one can say about such people is that they have enormous potential for growth.

If I did not love my profession and admire its practitioners I would not judge them—us—so severely.

The neo-conservative mood of the moment, as I said at the outset of this work, suggests that in the civilizing of engineers things may get worse before they get better. Yet the ever-improving caliber of today's engineering students gives reason for optimism. And the present state of our society lends new urgency to the need for a humanistic renaissance in engineering.

The comparison of the United States to ancient Rome has been made by many thoughtful people and noted in these pages as well. The words with which Edith Hamilton concluded her classic, *The Roman Way,* seem particularly apt:

> The old virtues were completely inadequate for the new day. The abilities of the pioneer and the conquerer, which had made the empire, could not meet the conditions which resulted from their achievements. To overcome nature or nations calls for one set of qualities; to use the victory as a basis for a better state in human affairs calls for another. When men must turn from extending their possessions to making wise use of them, audacity, self-reliance, endurance, are not enough. . . . It is worth our while to perceive that the final reason for Rome's defeat was the failure of mind and spirit to rise to a new and great opportunity, to meet the challenge of new and great events. Material development outstripped human development; the Dark Ages took possession of Europe and classical antiquity ended.[8]

I once entitled a speech "Technical Training and Liberal Learning: The Odd Couple." My reference, of course, was to the famous Odd Couple of stage, screen, and television: Oscar Madison, the lovable, sloppy sports writer, and his roommate, Felix Unger, the compulsively neat photographer. In the never-never land of television reruns, actors Jack Klugman and Tony Randall have inscribed these two characters into American folklore. Oscar and Felix represent two sides of our own human nature, carried to humorous extremes. Totally incompatible, they nevertheless live together in chaotic harmony. They understand each other, forgive each other, and—in spite of their mutual exasperation—seem to need each other, and this is what makes their relationship so heartwarming.

In preparing my speech, I intended to compare Oscar and Felix to engineering and the liberal arts, different but complementing one another, each compensating for the other's shortcomings. But the analogy doesn't fit very well. It is not even clear which character might be identified with which academic field. Should we make Felix, the fastidious intellectual, stand for the liberal arts, and Oscar, the earthier of the two, be engineering? Or, more likely, the other way around: Felix's precision going with engineering, and Oscar's loose and easy ways standing for the arts.

Even though the analogy didn't work the way I hoped it might, in considering the Odd Couple metaphor I came upon a reassuring message, a message that may appear to have little to do with engineering but that speaks to the point at which we have arrived: the possibility of change.

In the television series, Oscar and Felix are ever the same, never varying in their quirkiness, never learning from their mistakes. But the original play by Neil Simon tells quite a different story. At the end of Act Two, an exasperated Oscar says to Felix, "You mean you're not going make any effort to change? This is the person you're going to be—until the day you die?" Felix replies glumly, "We are what we are." Near the end of the third act, however, Felix reveals that he expects to get back in touch with his estranged wife and—in a remark that is crucial to the meaning of the play—says, "I'm not the same man she kicked out three weeks ago." And in the final lines, just before the curtain falls, Oscar, who started out as the quintessential slob, says to his pals, "Then let's play poker, and watch your cigarettes, will you? This is my house, not a pigsty."

In the world of television reruns Oscar and Felix are trapped forever in their stereotypes. But in the ending of the original play we are offered the possibility of change and redemption. Needless to say, I prefer the original.

23

Thoughts from the Dais: A Personal Coda

What am I doing here? Why am I seated on the dais in a hall where I have never been before, looking out over a dinner crowd of four hundred people, not one of whom I had ever met when I arrived here an hour ago?

To my left sits a young woman, the wife of the president of the alumni association, and beyond her the president of the institute itself. Yes, this is a respected engineering college, that much I know, and we are present at the annual alumni banquet. To my right sits the wife of the other honoree. Ah, yes, the other honoree. I am here to receive an award of honor.

Four months ago the letter arrived:

> Each year the Alumni Association of the Institute honors one or more outstanding leaders in science, engineering, or government by conferring upon those duly selected the Honor Award and Medal. . . . We would be pleased if you would allow us to bestow the Honor Award upon you this year. Presentation of the Award would be made at our Annual Alumni Banquet.

Then came the follow-up telephone calls. Would I accept? Yes, with pleasure. Would my wife come with me? I'm sure she'd be delighted. Would I please send in a brief biography and a photo for the alumni newsletter? Of course. Is this the way you would like your name engraved on the medallion? Fine. One more question. Would you be willing to make a few remarks to the assembled guests? Just five or ten minutes. Well, I guess so.

The weeks passed. All of a sudden it was yesterday and I was struggling to set down in writing a few words from the hundred speeches I had given in my head since the arrival of the invitation. Now, I feel a fleeting panic and quickly reach inside the breast pocket of my jacket. The folded papers are still there. My remarks. Will they be too long or too short, too ponderous, too flippant? What do these people expect?

Just an hour ago my wife and I drove onto the campus, were greeted by a waiting escort, ushered into a large reception room, and handed drinks. The president and his wife came over to meet us, affable and solicitous. We talked about cats and the cost of higher education. Then there were other hands to shake, pleasantries to exchange, names to repeat and forget. Just before the designated dinner hour, the director of alumni affairs descended upon us, accompanied by a photographer and assistants. I found myself posing for pictures, smiling awkwardly into the middle distance somewhere between the eyes of my fellow posers and the eye of the camera. Then in a flurry of excitement the dignitaries were lined up, and we paraded—there is no other word for it—through the dining hall, amongst the assembled crowd, and up onto the dais. I had barely sat down when we were all on our feet singing "The Star-Spangled Banner." A clergyman gave an invocation, someone introduced each of us at the head table, and we were finally free to start in on the melon.

What am I doing here? I lean back and look down to the other end of the dais, catching my wife's eye. She smiles wanly as if to say, "Why aren't we home, or in a little French restaurant, or at the movies?" Then I see her in conversation with my fellow honoree, a distinguished engineer, vice president of a major oil company, with a list of degrees, honors, and inventions as long as the table. His wife, to my right, asks me about my children.

She speaks with a graciousness compounded of executive-suite self-assurance and a delectable Louisiana accent. Her husband is an important man who has contributed much to our profession, and also happens to be a trustee of this institution. But why am I here?

I am here because I have written a couple of books and a bunch of articles that treat engineering from a "philosophical" point of view. Someone on the award selection committee must have read some of my work and liked it. Very nice. But—let us face facts—that is not the *real* reason.

To uncover the real reason, we must start with the committee that plans this dinner each year and tries to make it a success. It is not easy to make much of an event out of an annual alumni banquet at an engineering college. Yet the committee must try, year after year. The alumni must be brought together for the sake of camaraderie, to help each other's careers, and, most importantly, to ensure continuing loyalty and financial support for alma mater.

Imagine the hapless committee, charged with planning the evening, trying to add some interest to the program. Who can they invite? Consider for a moment the mathematics of the situation. There are more than two hundred and fifty accredited engineering colleges, each one with its dinners, seminars, and other assorted festivities. There are engineering clubs in practically every city. There are about fifty national engineering societies, most with numerous committees and technical subdivisions, and all with local chapters across the country. The American Society of Civil Engineers alone, with only 100,000 members, supports more than 400 national committees, and more than 300 regional sections and branches, most with technical groups and committees of their own. There are at least 5,000 such engineering units in the nation, some of them meeting each month. There must be at least 10,000 engineering meetings each year, many of them featuring speakers, guests, moderators, panelists, and honorees. Now, consider that there are no more than

600,000 engineers in the whole country who are members of professional societies (out of a total of 1.6 million engineers); in other words, 60 per scheduled meeting. This means that, with an average of one speaker per event, any affiliated engineer, over a thirty-year period, stands a 50 percent chance of being called upon! And this does not even take into account all the charitable, religious, political, and social affairs in which engineers participate.

So the committee has a problem. The few famous engineers are overinvited, overhonored, and overexposed. And since most of the merely distinguished engineers are not exactly show-biz attractions, the committee's problem becomes more acute. I have been on such committees myself, and I know what they are like.

I begin to feel angry, realizing that I am here because of a timeworn, almost meaningless custom. Some member of the planning committee said, "Let's invite this guy who wrote a book," and everyone else thought, "Terrific! One less problem to worry about." I am a fund-raising gimmick, a necessary component of dinner, like the melon being removed and the steak being served.

The steak, as it turns out, is delicious, but I continue to brood. Soon dessert and coffee are served, and once again I check the papers in my pocket. But my turn is not yet. Not nearly. The president of the alumni association is at the microphone making jovial remarks. Then the president of the institute talks about rising costs, nearly balanced budgets, high morale, and firm resolve.

Now it is time for the awards. But not mine. First the Fiftieth Anniversary Awards for members of the appropriate class. Next an award for the past president of the alumni association. Then certificates of appreciation for six of the institute's trustees. The outpouring of praise and gratitude swells to a crescendo as the Alumni Award is given to a man who has been in the institute's administration for over forty years.

Finally, the honor awards. The other honoree receives his first. Now I am standing, listening to the awards chairman tell all these strangers who I am, where I went to school. . . . I am being handed a silver medallion, and I hear applause. It is time for my remarks.

I step up to the lectern and haltingly begin to speak, about engineering, about what the profession means to those of us who are engineers, and what I think it means to mankind. I look out across the host of unfamiliar faces. They are politely attentive. One elderly man appears to be smiling, and—is it my imagination?—nods encouragingly. "You are among friends," I imagine him saying. "Anyone who cares for our profession is welcome in this place." *Our* profession. Of course. I am no stranger here.

Suddenly I am filled with a feeling of warm comradeship. The anger of a few minutes ago has vanished. So what if the proceedings are trite and a little silly? We are here to express allegiance to our profession, our respect for learning, and our commitment to constructive works. How else, after all, are we to demonstrate to each other our regard for shared values except through rituals such as this banquet? Of course the custom becomes stale. But would it be better if we had no banquets and awarded no prizes?

I have concluded my remarks, and my fellow engineers have applauded again. Thank you, my friends. We are standing, singing the school song. The words are printed in the banquet program, and by the second verse I find myself joining in.

Though the years our paths may sever
And best of friends may part
We'll ne'er forget fond mem'ries
Treasured within the heart.

The moment is as ridiculous as the song, and as wonderful.

Acknowledgments

I am grateful for editors: Tom Dunne, who for the third time has seen me through from concept to manuscript to book (Any man of letters who fancies the technical parts of *Moby Dick* is an ideal editor for an engineer!); Lewis Lapham of *Harper's,* for whom I wrote a dozen articles in the late seventies and who did his best to sharpen both my wit and my style; Steven Marcus, who first asked me to write for *Technology Review,* and his successors who have given me the unique blend of freedom and discipline that goes with a regular column; Joseph Epstein of *The American Scholar,* whose kind invitation elicited the essay that eventually became Chapter 1 of this book; Bob Asahina, who when he was with *The New York Times Book Review* gave me several books to read and report on, which I enjoyed doing (particularly Tracy Kidder's *The Soul of a New Machine*—one of the best books about engineers there is); and all the other editors of magazines and journals with whom I have worked.

I am grateful for family. Judy, my wife, does all good things, the least of which is reading what I write and making sensible comments. My son, David, contributed a lot to this book. He reads widely in the fields of technology, politics, and law, and brings to my attention much of importance that I would never find on my own. He edited the manuscript deftly, helped clarify my thoughts, and saved me from many excesses of style and emphasis. Had I taken all his advice, this would have been a better book, but it never would have been published on schedule. My son, Jonathan, although now a Bostonian, continues to provide amiable moral support. Mail addressed to him—from every environmental, wildlife, holistic-oriented organization in the land—still pours into our New York home, a chastening reminder of his thoughtful reservations about technology.

I am thankful for Ginny. For many years, Virginia Crowley has typed all my articles and manuscripts, tracked down needed information and given invaluable help in many ways, all the while running the office with aplomb at Kreisler Borg Florman Construction Company. She even managed to keep her sense of humor when a power outage dropped the entire book manuscript off the word processor disk (Ah, technology!).

I am grateful for engineering schools and engineering societies, particularly those that invite me to interesting meetings and seminars. My special thanks to the American Society of Mechanical Engineers, Clarkson University, Manhattan College, the Stevens Institute of Technology (the "respected engineering college" of Chapter 23), and the Thayer School of Engineering at Dartmouth College for their recognition of my work—it is their way, I know, of expressing support for philosophical introspection among engineers.

As I made clear in Chapter 1, I owe a great debt to the Fieldston School (particularly Elbert Lenrow's senior seminar), and to the Ethical Culture Schools of which Fieldston is a part. My appreciation of Dartmouth and Columbia is also expressed in that chapter.

Thanks to Lynn Chu of the Glen Hartley Agency for reading the manuscript and helping improve it. Thanks also to Ellis Mount and Preston Thayer for carrying out research assignments.

Thanks finally to Samuel Johnson, the eighteenth-century sage who provides an apt quote for every occasion, and who now reminds me there are buildings to be built and things I should be doing at the office. "Life is not long," he wrote in a letter to Boswell, "and too much of it must not pass in idle deliberation how it shall be spent."

Notes

Introduction

1. Report by the Office of Technology Assessment summarized in "Panel Reports Sex Disparity in Engineering," *The New York Times* (December 26, 1985). Also poll conducted by the Cooperative Institutional Research Program reported in "Career Choices Favor Engineering Above Others," *ASCE News* (May 1985).

2. Mary Komarnicki and John Doble, "Public Attitudes Toward Engineering and Technology," a Public Agenda Foundation report to the National Academy of Engineering Office of Public Awareness (Washington, D.C.: 1986).

3. Data from 192,248 entering freshmen at 368 colleges and universities in 1981. Cooperative Institutional Research Program, Graduate School of Education, University of California, Los Angeles. Also 1984 figures as reported in "Survey Finds Materialism Rising Among College Freshmen," *The New York Times,* January 14, 1985.

Chapter 1

1. Robert Louis Stevenson, *Across the Plains* (New York: Charles Scribner's Sons, 1898), 49. Readers of *The Existential Pleasures of Engineering* will recognize the quote which I used as the title of a chapter.

2. Mark Van Doren, *Liberal Education* (Boston: Beacon Press, 1959), 16.

Chapter 2

1. R. P. Loomba, *An Examination of the Engineering Profession* (San Jose State College, Manpower Research Group, Center for Interdisciplinary Studies, 1968), 24.

2. Jerald M. Henderson, Laura E. Bellman, Burford J. Furman, "A Case for Teaching Engineers With Cases," *Engineering Education* (January 1983), 288. Also see *Technology Review* (August/September 1986), 14, for comments by the director of placement at M.I.T.: "They are attracted to the fast track and the front office."

3. See Note 3, Introduction.

4. Walter H. Griggs and Susan L. Manning, "Money Isn't the Best Tool for Motivating Technical Professionals," *Personnel Administrator* (June 1985), 63.

5. Louis Harris and Associates, Inc., "Electrotechnology and the Engineer," Study No. 832048 (January–February, 1984), 12.

6. Friedrich Klemm, *A History of Western Technology* (New York: Charles Scribner's Sons, 1959), 343.

7. George Santayana in *The Great Ages of Western Philosophy, Volume VI, The Age of Analysis* (Boston: Houghton Mifflin Company, 1957), 63.

8. Griggs and Manning, 74.

Chapter 3

1. George M. Trevelyan, "Clio Rediscovered," in Fritz Stern (ed.), *The Varieties of History* (Cleveland: The World Publishing Company, 1956), 236.

2. Quoted in Benjamin Farrington, *Greek Science* (Baltimore: Penguin Books, 1953), 12.

3. Translation by R. C. Jebb in Whitney J. Oates and Eugene O'Neill, Jr., editors, *The Complete Greek Drama* (New York: Random House, 1938), Volume I, 432.

4. Farrington, 28.

5. Quoted in Friedrich Klemm, *A History of Western Technology* (New York: Charles Scribner's Sons, 1959), 22. This remarkable book consists mainly of excerpts from original historical sources, and I never tire of browsing through it.

6. National Research Council, Committee on the Education and Utilization of the Engineer, *Engineering Education and Practice in the United States* (Washington, D.C.: National Academy Press, 1985), 74.

7. Klemm, 23.

8. Klemm, 129.

Chapter 4

1. Quoted in Hans Straub, *A History of Civil Engineering* (London: Leonard Hill Limited, 1952), 114.

2. Klemm (see Note 5, Chapter 5), 262.

3. Straub, 117.

4. Ibid.

5. Quoted in Daniel H. Calhoun, *The American Civil Engineer, Origins and Conflict* (Cambridge: The Technology Press, 1960), 196.

6. Quoted in James Kip Finch, *Engineering and Western Civilization* (New York: McGraw-Hill Book Company, Inc., 1951), 92.

7. See particularly the writings of Derek J. deSolla Price, for example, "Of Sealing Wax and String," *Natural History* (January 1984), 49–56.

8. Klemm, 225.

9. Ibid., 240.

10. Quoted in W. H. G. Armytage, *A Social History of Engineering* (Cambridge: The M.I.T. Press, 1961), 94.

11. Ibid., 185.

12. Calhoun, 12.

13. Ibid., 27.

14. Quoted in William E. Wickenden, *A Comparative Study of Engineering Education in the United States and in Europe: Bulletin Number 16 of the Investigation of Engineering Education* (The Society for the Promotion of Engineering Education, 1929), 60.

15. Charles Riborg Mann, *A Study of Engineering Education* (New York: The Carnegie Foundation, 1918), 16.

16. Raymond H. Merritt, *Engineering in American Society, 1850–1875* (Lexington: The University Press of Kentucky, 1969), 118.

17. L.T.C. Rolt, *Victorian Engineering* (Harmondsworth: Penguin Books, 1970), 161.

18. William C. Wickenden, *Report of the Investigation of Engineering Education, 1923–1929* (Pittsburgh: Society for the Promotion of Engineering Education, 1930), 232.

19. David F. Noble, *America by Design* (New York: Alfred A. Knopf, 1977), 39.

Other principal references for Chapters 3 and 4:

Eric Ashby, *Technology and the Academics* (London: Macmillan & Co., Ltd, 1963).

Donald Hill, *A History of Engineering in Classical and Medieval Times* (LaSalle, Ill.: Open Court Publishing Company, 1984).

Thomas Parke Hughes (ed.), *The Development of Western Technology Since 1500* (New York: Macmillan Company, 1964).

Melvin Kranzberg and Carroll W. Pursell, Jr. (eds.) *Technology in Western Civilization* (New York: Oxford University Press, 1967).

Elting E. Morison, *From Know-How to Nowhere* (New York: Basic Books, Inc., 1974).

Kenneth P. Oakley, *Man the Tool-Maker* (Chicago: The University of Chicago Press, 1957).

Carroll W. Pursell, Jr. (ed.), *Technology in America* (Cambridge: The M.I.T. Press, 1981).

Anthony F. C. Wallace, *The Social Context of Innovation* (Princeton: Princeton University Press, 1982).

Chapter 5

1. National Research Council (see Note 6, Chapter 3), 36.

2. Ibid., 35.

3. Finch (see Note 6, Chapter 4), 52.

4. National Research Council, 92.

5. *Engineering Manpower Bulletin* published by the Engineering Manpower Commission of American Association of Engineering Societies, Inc., Number 80 (May 1986).

6. National Research Council, 91.

7. James B. Owens, quoted in "AAES' Leaders Revamp Budget, Agree on Goals," IEEE *The Institute* (October 1984).

8. J. Bronowski, *Science and Human Values* (New York: Harper & Row, 1956), 80.

9. Bertrand Russell, *Mysticism and Logic* (New York: Doubleday Anchor Books, 1957), 29.

10. Theodore Roszak, *Where the Wasteland Ends* (New York: Doubleday & Company, Inc., 1972), Chapter 5.

11. George Santayana, *Scepticism and Animal Faith* (New York: Dover Press, 1923), vi.

Chapter 6

1. Rustum Roy, "Optimum Technology," *Bulletin of Science, Technology & Society,* Volume 4, Number 4 (1984).

2. H. C. Luegenbiehl, "Society as Engineers' Client," *The Liberal Studies Educator,* Clarkson College, Volume 4 (1981–82).

3. James Boswell, *The Life of Samuel Johnson* (1791), record of April 7, 1775.

4. William James, *Pragmatism* (New York: Meridian Books, Inc., 1955), 18.

5. Ibid., 19.

6. "Proposed AAES Ethics Code Survey Results Reported," IEEE *The Institute* (November 1983), 9.

7. Stephen H. Unger, "The AAES Model Ethics Code," *IEEE Technology and Society Magazine* (June 1986), 2.

8. Philip M. Kohn and Roy V. Hughson, "Perplexing Problems in Engineering Ethics," *Chemical Engineering* (May 5, 1980), 101–107, and (September 22, 1980), 132–147.

9. "ASCE News," *Civil Engineering* (April 1984), 78.

Chapter 7

1. "Professionally Speaking," *Civil Engineering* (February 1982), 104.

2. John G. Burke, "Technology and Government," *Technology and Social Change in America* (New York: Harper & Row, 1973), 109.

3. John Kolb and Steven S. Ross, *Product Safety & Liability: A Desk Reference* (New York: McGraw-Hill, 1980), reviewed and quoted by J. T. Kane in *Engineering Times* (July 1980).

4. Jaclyn Fierman, "Why Enrollment Is Up at Quality College," *Fortune* (April 29, 1985), 170.

5. "Tip of the Liability Iceberg," *Engineering News-Record* (July 25, 1985), 9.

6. "Design Liability Rates Soar to New Record," *Engineering News-Record* (October 2, 1986), 12.

7. *Fortune* (April 29, 1985), 170.

8. National Research Council (see Note 6, Chapter 3), 105.

9. "Readers Express Mixed Attitudes on Personal Involvement in Defense Work," IEEE *The Institute* (February 1984), 3.

10. "Society Questions Balance of IEEE Aerospace Meeting," IEEE *The Institute* (January 1986), 1.

11. Irving R. Kaufman, "Charting A Judicial Pedigree," *The New York Times* (January 24, 1981), 23.

12. Robert J. Bork, "Morality and the Judge," speech excerpt, *Harper's* (May 1985), 29.

13. "Working Profile: Justice John Paul Stevens," *The New York Times* (July 23, 1984).

14. Kaufman, 23.

15. Irving R. Kaufman, "To Keep Lawyers from Going Wrong," *The New York Times* (March 26, 1985).

Chapter 8

1. Edwin T. Layton, Jr. "Engineering Needs a Loyal Opposition: an Essay Review," *Business & Professional Ethics Journal* (Spring 1983), 51–59.

2. Neal FitzSimons, Engineering Counsel, Kensington, Maryland, telephone conversation, 1985.

3. Miroslav Matousak, *Outcome of a Survey on 800 Construction Failures,* Swiss Federal Institute of Technology, Zurich, 1977.

4. John W. Gardner, *Excellence* (New York: Harper & Row, 1961), 159.

5. See my *The Existential Pleasures of Engineering* (New York: St. Martin's Press, 1976), Chapter 3.

6. William W. Lowrance, *Modern Science and Human Values* (New York: Oxford University Press, 1985), Chapter 4.

7. National Research Council (see Note 6, Chapter 3), 47.

8. See *History of Public Works in the United States 1776–1976* (Chicago: American Public Works Association, 1976), and other publications by the APWA.

Chapter 9

1. Josiah Royce, *The Philosophy of Loyalty* (New York: Macmillan Company, 1908), 118.

Chapter 10

1. Tekla S. Perry "How to Make It Big: Engineers as Entrepreneurs," *IEEE Spectrum* (July, 1982), 55.

2. Editorial, *Engineering News* (July 5, 1906).

3. Wickenden (see Note 18, Chapter 4), 232.

4. National Research Council (see Note 6, Chapter 3), "Engineering Employment Characteristics," sub-report by Panel on Engineering Employment Characteristics, 21.

5. Quoted in Carol Truxal, "EE Careers: New Directions But Old Issues," *IEEE Spectrum* (June 1984), 56.

Chapter 11

1. Lewis Thomas, "On My Magical Metronome," *Discover* (January 1983), 59.

2. Leon E. Trachtman, "The Public Understanding of Science Effort: A Critique," *Science, Technology, & Human Values* (Summer 1981), 10–15.

Chapter 16

1. *Congressional Record—House* (May 14, 1980), H3649.

2. All quotes and references from the hearings are recorded in *Comparative Risk Assessment,* Hearings before the Subcommittee on Science, Research and Technology of the Committee on Science and Technology, U. S. House of Representatives, 96th Congress, Second Session (May 14, 15, 1980), No. 129.

3. *American Textile Manufacturers Institute, Inc. v. Donovan,* 452 U.S. 490, 546 (1981) (Rehnquist, J., dissenting).

Chapter 17

1. Herbert Hoover, *Memoirs of Herbert Hoover* (New York: Macmillan Company, 1951), Chapter 11 of Volume I.

2. "Truly Spoken: An Admiral Sets NASA Straight," *Time* (April 7, 1986), 26.

3. "The Frailties of Machines and Men," Editorial, *The New York Times* (March 2, 1986).

4. Don K. Price, *The Scientific Estate* (Cambridge: Harvard University Press, 1965), 195.

Chapter 18

1. Herbert Hoover (see Note 1, Chapter 17).

2. National Research Council (see Note 6, Chapter 3).

3. *A Profile of the Engineer: A Comprehensive Study of Research Relating to the Engineer,* prepared by the Research Department of Deutsch & Shea, Inc., issued October 1957 by Industrial Relations Newsletter, Inc.

4. Harold Bell Wright, *The Winning of Barbara Worth* (New York: A. L. Burt Company, 1911), 87.

5. A. T. Robinson, "Culture and Engineering," *School and Society* (July 22, 1916), 144.

6. John S. Peck, "Does the Engineer Need Culture?" *American Association of University Professors Bulletin* (March 1956), 52.

7. Tom Wolfe, "The Tinkerings of Robert Noyce," *Esquire* (December 1983), 372.

8. Ibid., 371.

9. "Survey Results," IEEE *Spectrum* (June 1984), 59–63. Also "Job Satisfaction Rises with Age, Survey Reports," IEEE *The Institute* (November 1985), 1.

10. S. Edward Warren, "On the Future Development of Scientific Education in America," *Journal of The Franklin Institute* (April 1868), 283.

11. David Starr Jordan, "University-Building," *The Popular Science Monthly* (August 1902), 63.

12. Charles P. Steinmetz, *Steinmetz the Philosopher* (Mohawk Development Services, Inc., 1965), 129.

13. Walter F. Williams, *Proceedings 52nd Annual Meeting, Accreditation Board for Engineering and Technology* (October 23–26, 1984), 10.

14. Donald A. Rikard, Op. Cit., 21.

15. Walter Vogel, Op. Cit., 48–49.

16. Cited in *Manpower Comments* (Commission on Professionals in Science and Technology, December 1985), 8.

17. Cited in "Academe, Industry, and Government Assess Future of Engineering Education," *Engineering Education News* (July–August 1985), 1.

18. Samuel C. Florman, *Engineering and the Liberal Arts* (New York: McGraw-Hill Book Company, 1968), 10–11.

19. Derek C. Bok, "Can Ethics be Taught?" *Change* (October 1976), 28.

20. James Reston, "Washington, The Dull But Hopeful World of the Future," *The New York Times* (December 13, 1964).

21. William Ruder, "A Public Relations Executive Looks at the Engineer and His Role in Today's Society," *American Engineer,* March, 1966.

22. National Research Council (see Note 6, Chapter 3), "Support Organizations for the Engineering Community," sub-report by Panel on Support Organizations for the Engineering Community, 66.

Notes 5, 6, 10 and 11 above, plus much other material of great interest, found in Edwin J. Holstein and Earl J. McGrath, *Liberal Education and Engineering* (New York: Teachers College, Columbia University, 1960).

Chapter 19

1. Information about Born, Heisenberg, and Pauli from Robert Jungh, *Brighter than A Thousand Suns* (New York: Harcourt, Brace and Company, 1958), 17, 25, 158.

2. Ibid., 43.

3. Ibid., 132.

4. Jeremy Bernstein, *Experiencing Science* (New York: Basic Books, Inc. 1978), 48, 52.

5. Robert S. Root-Bernstein, quoted in "How Creative People Have Had Their Lives Changed by Big Prizes," *The Wall Street Journal* (June 17, 1985), 1.

6. Brooke Hindle, *Emulation and Invention* (New York: New York University Press, 1981), 142.

7. Albert W. Smith, *John Edson Sweet* (New York: American Society of Mechanical Engineers, 1925), 6.

8. Thomas P. Hughes, *Networks of Power* (Baltimore: The Johns Hopkins University Press, 1983), 113.

9. Louis Untermeyer, *Makers of the Modern World* (New York: Simon & Schuster, 1953). Some historians of technology do not share Untermeyer's certainty about Edison's interest in reading.

10. David McCullough, *The Great Bridge* (New York: Simon and Schuster, 1972).

11. Hughes, 453.

12. Hughes, 339.

13. "ET Interview," *Engineering Times* (February 1984), 16.

14. Vannevar Bush, *Pieces of the Action* (New York: William Morrow and Company, Inc., 1970), 268, 270.

15. Jeremy Bernstein, *Three Degrees Above Zero: Bell Labs in the Information Age* (New York: Charles Scribner's Sons, 1984), 49–69.

16. William J. Broad, "The Secret Behind Star Wars," *The New York Times Magazine* (August 11, 1985), 32–51. Also "A Laser's Inventor Ending Arms Work," *The New York Times* (September 11, 1986).

17. "Playboy Interview," *Playboy* (February 1985), 49.

18. Peter Petre, "America's Most Successful Entrepreneur," *Fortune* (October 27, 1986), 24.

19. Michael Schrage, "Alan Kay's Magical Mystery Tour," *TWA Ambassador* (January 1984), 33.

20. David McCullough, "Civil Engineers Are People," *Civil Engineering* (December 1978), 46–50.

Chapter 20

1. Wickenden (see Note 18, Chapter 4), 148.

2. James Kip Finch quoted in Eugene S. Ferguson, "Surveys of Engineering Education, 1918–1982," speech prepared for delivery at National Research Council seminar, Washington, D.C. (July 1983), 7.

3. Ferguson, 7.

4. Alfred P. Sloan Foundation, "An Evaluation of the Foundation's Program in the Social Dimensions of Engineering Practice" (October 3, 1980).

5. Letter from Alfred P. Sloan Foundation to presidents of thirty liberal arts colleges (March 5, 1982).

6. National Research Council (see Note 6, Chapter 3), iv.

7. Deans quoted in "Five-Year Program: A Great Idea?" *Engineering Times* (December 1985), 6.

8. "Wave of Curriculum Change Sweeping American Colleges," *The New York Times* (March 10, 1985), 1.

9. E. D. Hirsch, Jr., "Cultural Literacy and the Schools," *The American Educator* (Summer 1985).

10. William J. Bennett, *To Reclaim a Legacy: A Report on the Humanities in Higher Education* (National Endowment for the Humanities, November 1984), 8.

11. National Research Council, 74.

Chapter 21

1. Vladimir Shlapentokh, "In Soviet, Women Emerge Superior," *The New York Times* (February 4, 1984), 23.

2. Mary Diederich Ott, "Experiences, Aspirations, and Attitudes of Male and Female Freshmen," *Engineering Education* (January 1978), 326–338.

3. Reported in Vivian Weil, Illinois Institute of Technology, "Character Traits, Ethics, and the Professions," transcript of Radio Address, PBS Chicago (March 1977), 10.

4. Carol J. Auster, Franklin and Marshall College, reported in *Engineering Education News* (July/August 1984), 3.

5. Diane Tarmy Rudnick, "Women and Men in Engineering Technology: Shaping the Future," *Engineering Education* (May 1984), 716–720.

6. Vivian Weil, 14.

7. Marilyn Ferguson, *The Aquarian Conspiracy* (New York: J.P. Tarcher, Inc. and St. Martin's Press, 1980), 228.

8. Sherry Turkle, "Women and Computer Programming, A Different Approach," *Technology Review* (November/December 1984), 49.

9. "De Beauvoir On the 'New Femininity,'" *Harper's* (July 1984), 20.

10. Stephen Jay Gould, "Science and Gender," *The New York Times Book Review* (August 12, 1984).

Chapter 22

1. Harry Levinson, "Criteria for Choosing Executives," *Harvard Business Review* (July–August 1983), 113.

2. Wickenden (see Note 18, Chapter 4), 129.

3. Stephen D. Bechtel, Jr., *U.S. Scientists and Engineers* (National Science Foundation, 1983).

4. National Research Council (see Note 6, Chapter 3), "Engineering Infrastructure Diagramming and Modeling," sub-report by Panel on Engineering Infrastructure, Diagramming and Modeling, 71.

5. Quoted in Donald Christiansen, "Is the Four-Year-Baccalaureate Obsolete?" IEEE *Spectrum* (August 1984), 29.

6. Gerard Swope, "The Engineer and Social Development," response to award of the Hoover Medal for 1942. Available in the Engineering Societies Library, New York City.

7. Steinmetz (see Note 12, Chapter 18), 123.

8. Edith Hamilton, *The Roman Way* (New York: Mentor Books, 1957), 115.

Index